HUMAN VARIATION AND NATURAL SELECTION

SOCIETY FOR THE STUDY OF HUMAN BIOLOGY

Although there are many scientific societies for the furtherance of the biological study of man as an individual, there has been no organization in Great Britain catering for those (such as physical anthropologists or human geneticists) concerned with the biology of human populations. The need for such an association was made clear at a Symposium at the Ciba Foundation in November 1957, on "The Scope of Physical Anthropology and Human Population Biology and their Place in Academic Studies". As a result the Society for the Study of Human Biology was founded on May 7th, 1958, at a meeting at the British Museum (Natural History).

The aims of the Society are to advance the study of the biology of human populations and of man as a species, in all its branches, particularly human variability, human genetics and evolution, human adaptability and ecology.

At present the Society holds two full-day meetings per year—a Symposium (usually in the autumn) on a particular theme with invited speakers, and a scientific meeting for proffered papers. The papers given at the Symposia are published and the monographs are available to members at reduced prices.

Persons are eligible for membership who work or who have worked in the field of human biology as defined in the aims of the Society. They must be proposed and seconded by members of the Society. The subscription is £3 per annum (this includes the Society's journal *Annals of Human Biology*) and there is no entrance fee.

Applications for membership should be made to Dr. A. J. Boyce, Hon. General Secretary, Department of Human Anatomy, University of Oxford, South Parks Road, Oxford OX1 3QX.

PUBLICATIONS OF THE SOCIETY

Symposia, Volume I, 1958: *The Scope of Physical Anthropology and its Place in Academic Studies,* edited by D. F. ROBERTS and J. S. WEINER (out of print).

Symposia, Volume II, 1959: *Natural Selection in Human Populations,* edited by D. F. ROBERTS and G. A. HARRISON. Pergamon Press (members £1).

Symposia, Volume III, 1960: *Human Growth,* edited by J. M. TANNER. Pergamon Press (members 53p).

Symposia, Volume IV, 1961: *Genetical Variation in Human Populations,* edited by G. A. HARRISON. Pergamon Press (members £1).

Symposia, Volume V, 1963: *Dental Anthropology,* edited by D. R. BROTHWELL. Pergamon Press (members £1.25).

Symposia, Volume VI, 1964: *Teaching and Research in Human Biology,* edited by G. A. HARRISON. Pergamon Press (members £1.25).

Symposia, Volume VII, 1965: *Human Body Composition, Approaches and Applications,* edited by J. BROZEK. Pergamon Press (members £3).

Symposia, Volume VIII, 1968: *The Skeletal Biology of Earlier Human Populations,* edited by D. R. BROTHWELL. Pergamon Press (members £2).

Symposia, Volume IX, 1969: *Human Ecology in the Tropics,* edited by J. P. GARLICK and R. W. J. KEAY. Pergamon Press (members £1.50).

Symposia, Volume X, 1971: *Biological Aspects of Demography,* edited by W. BRASS. Taylor & Francis (members £1.90).

Symposia Volume XI, 1973: *Human Evolution,* edited by M. H. DAY. Taylor & Francis (members £1.90).

Symposia, Volume XII, 1973: *Genetic Variation in Britain,* edited by D. F. ROBERTS and E. SUNDERLAND. Taylor & Francis (members £2.65).

Symposia Volume XIII, 1975: *Human Variation and Natural Selection,* edited by D. F. ROBERTS. Taylor & Francis (members £2.55).

SYMPOSIA OF THE
SOCIETY FOR THE STUDY OF HUMAN BIOLOGY

Volume XIII

HUMAN VARIATION
AND NATURAL SELECTION

Edited by

D. F. ROBERTS

TAYLOR & FRANCIS LTD
LONDON

HALSTED PRESS
(a division of John Wiley & Sons Inc.)
NEW YORK–TORONTO
1975

First published 1975 by Taylor & Francis Ltd, London, and
Halsted Press (a division of John Wiley & Sons Inc.), New York

© 1975 Taylor & Francis Ltd

Printed and bound in Great Britain by Taylor & Francis (Printers) Ltd,
Rankine Road, Basingstoke, Hampshire.

Taylor & Francis   ISBN 0 85066 080 7

Library of Congress Cataloging in Publication Data

Main entry under title:
Human variation and natural selection.
    (Symposia of the Society for the Study of Human Biology; v. 13)
    Reprint of 2 papers from The scope of physical anthropology and its
place in academic studies, edited by D. F. Roberts and J. S. Weiner, issued
as v. 1 of Symposia of the Society for the Study of Human Biology; of
Natural selection in human populations, edited by D. F. Roberts and G. A.
Harrison, issued as v. 2 of the series; and of Genetical variation in human
populations, edited by G. A. Harrison, issued as v. 4 of the series.
    Includes bibliographies and indexes.
    1. Human population genetics. 2. Variation (Biology). 3. Natural
selection. I. Roberts, Derek Frank. II. Roberts, Derek Frank, ed. The
scope of physical anthropology and its place in academic studies. III.
Roberts, Derek Frank, ed. Natural selection in human populations. 1975.
IV. Harrison, Geoffrey Ainsworth, ed. Genetical variation in human populations. 1975.   V. Series: Society for the Study of Human Biology. Symposia;
v. 13. [DNLM: 1. Selection—Congresses. 2. Variation—Congresses.
W3 SO5915 v. 13/QH375 H918]
QH431.H839   573.2′1′3   75–12979
ISBN 0–470–72590–7

PREFACE

THE early volumes of the symposium series of the Society for the Study of Human Biology have been out of print and unobtainable for several years. Many of the papers they include were remarkably forward-looking and are of as much interest today as they were when first presented, others are still valuable as simple introductions to topics about which now much more is known. The Society therefore considered it desirable to reprint these early studies; the present volume is devoted to the papers on human genetics that appeared in volume 2 "Natural selection in human populations" and volume 4 "Genetical variation in human populations", and with these are included two genetic papers from volume 1 "The scope of physical anthropology and its place in academic studies". Most of the other papers in volume 1 were updated and included in the later symposium on "Teaching and research in human biology" (volume 6). In the present reprinting, each paper is accompanied by a brief updating paragraph, commenting on it in the light of developments in the intervening years. These comments are not intended to be comprehensive, but merely to be a guide to some of the recent trends in understanding of each topic, with a few key references.

This reprint the Committee would like to associate with the name of the late Professor L. S. Penrose (1898–1972). He was a good friend to the Society, particularly in its infancy. He participated with enthusiasm in the Symposium at the CIBA Foundation at which the Society was born and previous to that, in the course of informal discussions, had much useful comment to make on the form of the Society and the scope of its possible activities. This volume is dedicated to the memory of Lionel Sharples Penrose, with gratitude for all that he did for human biology in general and the Society in particular.

Department of Human Genetics,
Newcastle   D. F. ROBERTS

v

APPRECIATION

LIONEL SHARPLES PENROSE (1898–1972), M.A., M.D., D.Sc., F.R.C.P., F.R.S.

IT is given to few men to influence the thinking of a generation. One such was Lionel Sharples Penrose, born in London on 11th June 1898, the son of the artist James Doyle Penrose. Educated at the Downs Preparatory School, Colwall, Leighton Park School, Reading, and St. John's College, Cambridge, he obtained a first class degree in moral sciences, despite interruption by a period of service with the Friends Ambulance Unit in France in 1918. Medicine attracted him, and after a year in the Psychological Department at Vienna University, he entered St. Thomas's Hospital London for his clinical studies, qualifying in 1928, and obtaining his M.D. in 1930.

His interests were formulated early. After a period of research in Cardiff City Mental Hospital, he was appointed a Medical Research Council medical officer at the Royal Eastern Counties Institution, Colchester, and in the nine years there his classic investigations into the causes of mental defects were prosecuted and the results published in an M.R.C. special report. His already great experience was extended by a period in Canada from 1939 to 1945, where he was Director of Psychiatric Research in Ontario, physician to the Ontario hospital, and lecturer in Psychiatry at the University of Western Ontario. In 1945 he was appointed Galton Professor of Eugenics at University College London, a post he held until he retired in 1965, when he became Director of the Kennedy Galton Centre, Harperbury Hospital, St. Albans.

In 1933 he took issue with the then prevalent idea that all mental defect was either due to injury or infectious disease or to the "neuropathic diathesis", and instead took the view, further developed in his research publications, that inborn mental defect was due to a very great variety of causes. This is now universally accepted. His work on phenylketonuria had considerable influence on the subsequent development of human biochemical genetics, without which the virtual elimination today of this disease in parts of Britain by screening procedures would not yet have been achieved. The reference tables he initiated for a variety of genetic problems have saved countless manhours of labour. The present advance in knowledge of genetic linkage in man stems directly from his pioneering interest, in the same way as developments in the last two decades in human cytogenetics, dermatoglyphics, and many other fields along the borderline of biology and genetics. The types of problem in which he was interested, and his broad biological attitude to them, meant that his writings offered much to those concerned with aspects of human biology other than inheritance. Indeed the debt to him of modern human biology as a whole is inestimable.

His work was recognised by many distinctions; the Buckstone Browne Medal in 1933, the Weldon Medal for biometrics 1950, F.R.S. 1953, F.R.C.P. 1962, Joseph P. Kennedy Jr. Foundation Award 1963, Leon Bernard Foundation Award for research into mental subnormality 1965, and honorary doctorates conferred by McGill University and the University of Newcastle upon Tyne.

His influence on the development of human genetics throughout the world was enormous, his views and attitudes being disseminated by his own students, by others whom he was always ready to help in personal discussion, by his many writings, and by his rigorous editorship of the *Annals of Human Genetics*. He did much to disseminate genetic knowledge amongst the lay public reaching a wide audience through broadcasting and television, largely because of the skill with which he presented information in a form easily understood by the average listener. His elementary student texts were models not only of clarity but of elegant English, and his book "The biology of mental defect" has become a classic. He served on innumerable committees, national and international. His personal influence was enhanced by his reticence, his sharp sense of humour and of the ridiculous, by his interesting mixture of enthusiasm and scepticism, and by the acuteness of his mind. He sought recreation in painting and chess, aptitudes for which were family traits, and, in the company of his wife who shared many of his professional and personal interests, in the appreciation of the English landscape and its richness.

C. A. Clarke

CONTENTS

HUMAN VARIABILITY AND ADAPTABILITY

L. S. PENROSE

In this short article, Professor Penrose set out with characteristic simplicity what seemed to him the most important fields of genetic investigation of relevance in human biology. Of the points made, the sixteen years since it was written have seen:

1. General acceptance of the importance of polymorphism as a major interest, and the emergence of a host of new problems—the very number of loci at which polymorphism occurs (Giblett 1969, Harris 1970), the re-emergence of arguments for neutral mutations as a non-Darwinian evolutionary force (King and Jukes 1969, Kimura 1968), the identification of the changes in the structure of the genes themselves that are responsible for polymorphic situations (e.g. Lehmann and Carrell 1969 for variant hemoglobins, Smithies et al 1966, Black and Dixon 1968 for haptoglobins), the associated problems of gene interaction, and the very recent field of chromosomal polymorphisms (Pearson 1973).

2. Recognition of the difficulties in quantifying the rate of gene frequency change, partly arising out of these developments.

3. Development of studies attempting to relate phenotypic variation to external conditions (e.g. Glanville 1969, Rimoin et al 1967, Steegman 1965, Roberts 1973).

4. Acceptance of the importance of gross abnormalities in population studies, for these were included in I.B.P. programmes and a major W.H.O. research effort was devoted to their study (Stevenson et al 1966).

THE polymorphic nature of human populations is the special feature which makes physical anthropology so fruitful. This polymorphism, or natural variability, is a consequence of genetical and physiological reaction to environment. The student of physical anthropology should be specially interested in the genes carried in the population he studies but the administrator, medical or industrial, finds that he cannot escape examining the individual variations, apparently caused by environment directly, which are largely independent of hereditary differences. For example, malnutrition alters body measurements and even changes pigmentation. Environmental conditions are continually changing and human populations are continually adapting themselves to existing conditions. There are two main aspects of this process to be studied and the two approaches have not yet been integrated: (i) The reaction of the human phenotype to its environment; (ii) The reaction of gene frequency to changes in fitness of the phenotype.

(i) We find that, in most of the classical traits used for defining races, such as weight, span, skin colour and even head measurements, the phenotype is considerably influenced by external conditions. Examples of traits measured in United States immigrants, by Boas and Lasker, showed that not only in the next generation after migration, but even in the immigrants themselves after a period of years, there are marked physical changes. Hygiene and diet not only influence growth but they also influence disease susceptibility. Consequently the study of physical measurements, whether it concerns stature, weight or fat, is a natural part of public health practice, or what is now called social medicine.

(ii) Characteristic of modern analysis is the attempt to estimate the direction and speed of gene changes. Natural selection is not a constant force; it is continually altering with changes in physical and social environments. Thus we must ascertain with the greatest possible precision the incidence of detectable phenotypes, from which gene frequencies can be fairly easily calculated. The selection pressures on these phenotypes can be gauged by studying their morbidity and mortality, a field which is only just beginning to be explored.

The reactions of genes to selective influences are often extremely complex as, for example, mere family size is a strong selective agent affecting the Rh locus and probably also the ABO locus. The smaller the family, the less danger is there of maternal immunization and consequent foetal damage. Other typical cases are thalassaemia and sickle cell trait where protozoal infection is significant, and cystic pancreatic fibrosis where selection has different force in different climates.

The rate of gene frequency change can be estimated by reference to methods developed by Fisher, Haldane and Wright. Genes either tend to become "fixed", that is universal or eliminated, or they persist in a state approximating to equilibrium; and unless the equilibrium is stable the gene is not likely to be permanent. The two methods of producing permanent polymorphism are (a) mutation balanced against selection, and (b) heterozygous advantage. Both (a) and (b) produce stable genic equilibrium—not merely equilibrium of the neutral type like that of Hardy and Weinberg. Changing levels of fitness of the genotypes can alter the stable equilibrium point, as possibly has happened in the case of the sickle cell trait in American Negroes. There will also be possible mutation rate changes to consider, remembering, for example, that doubling the mutation rate may eventually double the incidence of rare abnormalities.

Local changes in gene and phenotype frequencies have caused isolated populations to differ markedly from one another in many respects and have produced so-called *races*. These are best studied in relation to the human race as a whole, or the human species as some prefer to call it. All groups deviate from the world mean in their own measurements and from the world distribution in their gene frequencies. In the interests of logical simplicity, the whole world population should be treated as one race unless and until this is proved a fallacious assumption. The understanding of the nature and causes of divergence from the human mean in a given population then becomes the fundamental problem of physical anthropology, equivalent to the study of human evolution in terms of genes. However, as Professor Le Gros Clark has said about the statistical phase in the development of anthropology, in their present heyday the geneticists will probably be, like their predecessors, overconfident.

In practice, surveys of genetical anomalies are required as well as surveys of so-called normal characters. There is, in fact, no absolute distinction between the two. On which side of the fence are harmless anomalies like taste deficiency and colour blindness (common) or pentosuria (very rare)? From the genetical point of view the distinction between the normal and the abnormal is a mistake and it can retard researches: the same gene can be advantageous or disadvantageous according to its setting.

We need to know much more about the incidence of gross abnormalities and they are easy to detect. Anencephaly, like other foetal defects, can be important for anthropology as well as in medicine. Its incidence is extremely uneven when different populations are compared and it is easy to identify at birth. Lethal traits, like phenylketonuria and amaurotic idiocy, also have uneven distributions, comparable with blood group and pigmentation differences. To obtain information about the incidence of rare phenotypes and subject it to genetical analysis will give a new lead to the study of natural selection in man and will furnish new data upon the effects of fresh mutations. For such investigations to be practicable, co-operation of public health authorities, obstetricians and pathologists with anatomists and biochemists is required. WHO and UNESCO should both be interested in helping to obtain basic data though in official figures there is no substitute for individual field work. A new and broader idea of human polymorphism will be required before anthropologists can play their full part in human race genetics, into which their subject must, in my view, be to a large extent inevitably transformed.

BIOLOGICAL STUDIES OF
SMALL COMMUNITIES

A. C. Stevenson

This is among the most important papers ever written on genetic studies of human populations. Here, the fundamental problems of human genetics are approached from the viewpoint of a clinical geneticist, and they illustrate a rarely recognised intrinsic difference. To the latter who deals with abnormal conditions, cases resulting from new mutations are by no means unusual; the more severe the condition, the more likely it is to be the result of a new mutation. Those who work with normal monogenic variants rarely see mutations and the contrast in emphasis is clearly shown by comparison with the next paper by Penrose.

But the importance of this paper lies in the fact that here is laid down the rationale of studies based on total ascertainment of a condition, which proved so profitable in the many investigations of Dr. Stevenson and his colleagues (e.g. Stevenson and Cheeseman 1956 on deaf mutism; Wells and Kerr 1965 in ichthyosis; Davison 1965 on epidermolysis bullosa). It was this type of study that illustrated so clearly the problem of genetic heterogeneity, in which a condition, that appears to be a clinical entity from the point of view of diagnosis and treatment, proves to be made up of a number of separate entities often under different modes of genetic control. Largely through studies of this type, more attention has been given to examining some of the theoretical concepts on which population genetic studies were previously based.

THE title of my contribution to these discussions covers a very wide field. I propose therefore to limit the types of studies which I discuss to those with which I am personally familiar, as one who, starting work in the quiet pastures of epidemiology, has wandered into the treacherous groves of population genetics. I shall therefore first discuss some of the factors which determine the objects and mould the form of such genetical inquiries. I shall then mention some of the kinds of investigations into the genetical structure of human populations which have been undertaken in recent years. I shall conclude by describing the steps in planning, carrying out and analysing the findings of a typical investigation which illustrates the practical and theoretical problems encountered and the limitations imposed by methods and by the basic population information which is available.

At present perhaps the most hopeful and most widely used approach to understanding the genetical structure of human populations is that of using as "markers" certain traits whose distribution and pattern can be identified and studied in a community. Subsequently the findings can be examined to see how far they agree with theoretical explanations, with experimental work on animals or, more commonly, with both. In other words, as in so much other research, we are forced to be opportunistic and to examine such small facets as we can of a larger problem, hoping that they will contribute towards establishing a more specific hypothesis and that in time a series of such inquiries will provide the material for an intellectual synthesis.

In recent years we have been impelled to try to measure forces which, not so very long ago, were mainly considered qualitatively and were sometimes little more than abstractions. Much of this has been due to worries concerning the effects of energy transfer radiations in causing transmissible damage to cells in the germ tract. Another influence has been the example of the more exact relationships established between segregating traits and conditions at the appropriate loci, notably in work on the blood groups. The magnificent tradition established in this country of studying the population pattern of newly-discovered blood groups has not only often vindicated brilliant theoretical predictions, but has proved of very great practical value in medicine, in human genetics and in anthropology. There is some slight danger however that because, as in the case of the blood group alleles, theoretical considerations have been vindicated, such largely mathematical conceptions about other aspects of population structure are too readily accepted. Perhaps I had better be a little more specific, in giving reasons why it is so necessary to test by every possible means some of the theoretical conceptions which until a few years ago were seriously questioned only by very few.

For a variety of reasons, all stemming from the need to start from a grossly oversimplified concept or model, theoretical and mathematical geneticists have tended to view a population as consisting of individuals predominantly and ideally homozygous at the great majority of loci. To some extent these ideas sprang from, and to some extent they appeared subsequently to be reinforced by, botanical observations, where inbreeding is the rule and outbreeding the exception relative to most higher creatures. True, it was recognised that there were exceptions. Multiple allelism was accepted and other phenomena such as heterotic effects were dealt with in theories, but it seems fair to say that the picture which seemed to be visualised by many of the most distinguished theoreticians was that at least at a preponderance of loci, individuals were homozygous, and that the effects of selection pressure were as a whole consistently to perpetuate or to restore homozygosity. Another oversimplification which has stemmed from theoretical considerations, far in advance of the data by which they could be tested, has been that which concerns our thinking about patterns of mating. Clearly the simplest way to start is to postulate completely random and non-assortative mating and then to attempt to make allowances for departures from such a situation. In effect, the final picture which tends to emerge is of random mating only modified by rather more marriage of close relatives than would occur by chance. Yet we know in man of quite strong correlations between husbands and wives, such as those in height and intelligence, and as very little effort indeed has been devoted to the study of such phenomena we may well expect to see more established and possibly to discern their effects on distribution of certain characters.

Somewhat similar ideas, tending towards depicting a predominantly ideal homozygote situation in populations, have emerged from experimental work, not only in plants but for example in drosophila and in mice. Naturally, in much experimental work, notably that involving the use of very large numbers for experiments on the induction of mutation, it is a great advantage to use standard strains about which as much as possible is known and which will on the whole "breed true". It may legitimately be suspected, however, that at times these stocks have artificially been built to be homozygous at as large a number of loci as is compatible with viability and fertility. Further, in order to keep a stock to a manageable size and in the process of identifying mutant expressions in each generation, a large number of the creatures are sacrificed, which may tend to the same direction. True, multiple allelism has been detected at a small number of loci in most species, and experimental geneticists sometimes refer to visible expressions of mutants being perhaps those of the most "extreme" mutant at a locus, so implying the same idea. Nevertheless, deductions from such artificial populations must be applied to human populations with great care. Certainly in such experimental animals, as in man, at most loci where a harmful mutation can occur there is no evidence that any other exists. Yet other alleles *may* exist. In some cases there is suggestive evidence, and in any event these few identifiable loci may or may not be representative of the much greater number of "silent" loci, not identifiable because no gross effect determining mutation occurs or has been identified.

Attempts to allow free breeding in captive populations are extremely difficult but it is interesting to see how closely some of the problems of study of wild populations, notably in ornithology, correspond to those in human populations.

I may have laboured these points and I would not want to go on record as questioning the contribution from these sources to ideas of population structure or even of their validity for much that we observe at the present time. It does seem important, however, to point out that they can be questioned. What is predicted as the effects of increased mutation in man from radiation or other mutagenic agents will depend quantitatively and qualitatively on to what extent these classical views of population structure are accepted. Indeed it is only possible to make the most elementary numerical estimate of the effects if the ideal homozygote model is accepted. This is a point which appears to have escaped the notice of not a few who have made rather elaborate calculations.

Fundamental genetic problems

It is convenient to outline the kinds of practical studies of the genetical structure of populations which can be made and which represent the scope of inquiries at the present time, not by listing specific projects but rather by listing the kinds of questions which seem to be being asked when the answers are sought by practical investigations in defined populations. Here are some of the questions which seem to be posed in these investigations : —

1. To what extent are variations in individuals in populations determined by hereditary and by environmental factors? In the medical field, what, for example, are the relative roles of inborn factors and precipitating factors in disorders such as cancer at a specific site, rheumatic fever and tuberculosis? An important aspect here is in the hereditary contribution relative to that of environment in embryonic and in foetal deaths.

2. What single mutant determined traits occur in populations? What are the individual frequencies and the range of such frequencies? How are such frequencies likely to be influenced by improvements in medical and social care?

3. At what rate are fresh mutations occurring at identifiable loci in a population? What is the range of rates between different loci? What would be the effect on the phenotype frequencies if these rates were raised?

4. To what extent are the frequencies of such traits in populations determined by mutation pressure and to what extent by selection pressure? To what extent and in what direction does selection pressure act on the homozygote and on the heterozygote?

5. What happens when new mutations are introduced into a population? How quickly and at what cost in human suffering are they eliminated?

6. What is the pattern of certain continuously distributed traits in a population? How far can we distinguish in some such traits multiple gene effects from multiple allelism, and from alteration by environmental factors or by a few modifying genes of the effects of a single mutant?

7. How complete is the manifestation of certain genotypes? How specific is phenotype to genotype? In other words, does degree or form of manifestation vary more within or between families, and what evidence is there, for example, that more than one genotype can determine identical, or at least at present indistinguishable, effects?

8. In the rare instances where we can distinguish the heterozygote and the two homozygotes, as in the cases of the blood group alleles and of a few harmful mutants, do the relative genotypic distributions fit the hypothesis that only the mutant homozygote is at a relative selective disadvantage in the population, or is there any evidence to support heterozygote advantage or disadvantage?

9. In respect of a mutant gene determining a recognisable trait, is there any evidence of linkage with any of the blood group or other loci which can be used as markers?

As will be clear, the emphasis is on unfavourable traits because they are relatively easy to identify and to classify by the large range of diagnostic procedures used in medicine. To me, as a physician, one of the most interesting side effects, in work which has involved seeing all of the "cases" of a condition in a population, is that so often the clinical and the genetical classifications do not correspond. Often we, interested in genetics, must wait for clinical or biochemical separations before we can go further. But genetics has much to contribute to the classification and diagnosis of disease, by separating out for further study cases of a clinically homogeneous group which is heterogeneous genetically.

Lessons from genetic analysis of a total population

Perhaps an example which illustrates as well as any some practical problems of the work, and the kind of information to be gained from the study of the complete ascertainment of a mendelian trait, is the study of deaf mutism. For reasons which will emerge, it is not at present possible to calculate gene frequencies with confidence nor to estimate mutation rates, but much other genetical information which can be derived from such studies is exemplified.

A study of this kind was carried out from my department some years ago in the population of Northern Ireland. It may be instructive just to go through the stages of the work. We wanted to make a complete ascertainment so that the complete pattern of distribution of the trait would emerge, not one distorted by selective failure to record, for example, cases in older age groups or "sporadic" cases who had no affected relatives. Then we had to define what we wanted to ascertain, so that if at all possible we, and not those reporting cases, would make the final decision as to who would be included. So we told those from whom we sought assistance that we wanted the names and addresses of all who were born totally or substantially deaf or who became deaf under the age of five years. We thought of everyone likely to know such persons. We consulted the records of four schools for the deaf, the main one having remarkable records going back to 1834 and the two other large ones had records going back to about 1860. We wrote to every doctor in practice in the province. We consulted particularly otologists and the records of otological and children's hospitals and departments. We had every possible co-operation from four religious bodies who organise spiritual and social services for the deaf. We consulted all school and child welfare medical officers, welfare officers, disablement officers of the Ministry of Labour, the register of mental defectives and so on.

Then came the task of visiting all these people and their families, of separating out those who had early *acquired* deafness on clinical grounds and on the grounds of histories received, checked whenever possible against records made when the cases were first seen. We were also able to exclude from the study a number of subjects suffering from specific syndromes, hereditary or otherwise, of which deafness was but a part, and some where the trait was plainly early-onset perception deafness starting in some members at least before the age of five years. We were also able to identify and to exclude a small number of conditions which were present at birth but not of genetical import, such as those due to maternal rubella, rhesus incompatibility and congenital syphilis.

In the rest, we took a careful family history, examined affected members clinically, and in some cases did pure tone audiometric and labyrinthine function tests on both affected and unaffected members of families. (In two instances we were able to trace affected members of families who had been admitted to the Ulster School for the Deaf as far back as 1840 and we identified members of a family first detected by Sir William Wilde in the 1850s.)

Analysis of the data showed, in agreement with long held views, that a majority of the cases and families indicated inheritance by a single autosomal recessive mechanism although there were certain peculiarities. There is a complication here in that a few cases may have been determined by dominant mutants but for the sake of simplicity I shall ignore this in my present account.

Over 11% of hearing parents who had deaf-born children in our families were found to be related, and over 6% of them were as closely related as full cousins. The offspring of the thirty-six consanguineous marriages of hearing parents segregated in the expected ratio of a mendelian recessive hypothesis (making appropriate allowance for mode of ascertainment). There was a failure, however, to reach expectation of the proportion of deaf in the offspring of the 273 matings of hearing parents who were not related. Further analysis showed that this failure was entirely due to an excess in each sibship size of sibships with only one affected. Our hypothesis to explain this was that these represented children whose auditory apparatus was damaged in utero or at birth and that they really corresponded to the non-hereditary group like those determined by rubella and by syphilis, although we did not know the cause. In other words they were phenocopies. It may be noted that comparison of these sporadic cases and definite hereditary cases by pure tone audiometry and by tests of labyrinthine function failed to show differences; nor were there differences in parental ages.

At this stage it may be noted that estimates of prevailing consanguinity marriage rates were available from previous investigations and that these observed consanguinity rates of hearing parents who had deaf children were very high. These rates were too high for the case frequencies. In recessive conditions obviously the less common the condition the higher the expected, and usually the observed, consanguinity rates in parents. An obvious explanation would be that the case frequency represented the *sum* of frequencies of those homozygous for independent recessive mutants individually sufficiently uncommon to determine that such a high proportion of those affected had related parents.

Deaf mutism is unique as a recessive trait in man in two very important respects. Most such traits known are rather rare and lethal and so the subjects seldom grow up and marry. Even more important, a very high proportion of the deaf-born marry deaf mutes (although the cause of the deaf mutism does not appear to influence selection of mates so that there are all combinations of marriages of hereditary deaf, phenocopies and acquired deaf).

We had 48 matings of two deaf-born subjects, most of whom were definitely hereditary deaf; 12 matings involving one deaf-born partner and a hearing partner and 42 involving a deaf-born and an acquired deaf partner. The two latter categories can reasonably be equated from a genetical point of view. Study of these matings seemed to confirm the suggestion from the high consanguinity rates that there were a number of independent recessives determining the deaf mutism in this population.

Further study by pure tone audiometry and by the rotation test of labyrinthine function of deaf-born partners who had had all hearing offspring failed to detect differences. Yet in most of the subjects examined both must be presumed to have been homozygous for different mutants. However, our diagnostic armamentarium is very weak and cannot detect where, on the perception tract, an interrupting lesion is located.

In a preliminary study it had been demonstrated that there was no evidence of a change in the frequency of deaf mutism in Ireland over the past hundred years. The data available were the deaf school records and the remarkable Irish Censuses, where from 1851 to 1911, questions were asked about the deaf-born and the blind in all households. Data from subsequent follow-ups by the dispensary doctors were also recorded in the reports.

A study of the relative fertility of the deaf-born showed that it was only about one-third that of hearing subjects in the community so that there was, if this had always been so, a heavy loss of mutants to the population in each generation—a loss so heavy that it could only be balanced by mutation rates far higher than any likely to occur, on the analogy of mutation rates at other loci in man and on the range of mutations at a locus in animals.

A possible explanation would be that the rate was maintained by slightly increased fertility of heterozygote carriers who might be very frequent in the population. The increase of fertility would need only to be very small—too small to detect for practical reasons.

In conversation with Professor Penrose, he made an ingenious suggestion that if such carriers had slightly lower intelligence on an average than that in the population, an undue proportion would fall into that intelligence level group which had a higher fertility under modern conditions. Following up this suggestion, Mr Harrison, the Educational Psychologist of Belfast Corporation, examined some seventy children whom we chose on the basis of relationship to hereditary deaf subjects, and who, on a recessive hypothesis were predominantly single or double heterozygotes for a deaf mutism mutant. The mean IQ from three separate intelligence tests indicated that they were from 7–10 points above the population average!

It is clearly of great interest to determine the number of independent mutants which are contributing to hereditary deaf mutism and the relative and total gene frequencies of these mutants.

Although the problem can be tackled in a variety of ways from data such as those available from the study described, no reasonable estimates can be made. It is possible to calculate a minimum number in two ways, by studying the matings involving deaf-born persons but here even a small contribution of dominant mutants could be very misleading; and also by assuming that all the mutants are of equal frequency.

Rough estimates made in these two ways from our data did not agree. However, it seems likely that there are not less than 4–6 mutants and it is difficult to escape the conclusion that at least one hearing person in three in the population is heterozygous for a recessive deaf mutism gene.

It appears to me that the lessons which emerge from this study as far as population genetics are concerned are these : —

1. That deaf mutism is an example of a trait where the genetical determining systems are complex. This would have been much more difficult to recognize, and even so crudely to define, were it not for the special circumstances of the selective matings of deaf mute subjects. It may be that other apparently simply determined traits are in fact equally complicated, but as there are no opportunities for observing the offspring of a variety of matings, there is no evidence to suggest a complex explanation.

2. That genetical analysis has demonstrated a heterogeneity which would not be suspected clinically.

3. That advances in clinical methods, especially if heterozygotes can be detected, would open up fascinating possibilities in the study of deaf mutism. That genetical analysis has pointed the way to some such clinical studies.

4. That even with the complexities of the genetical situation, unless the estimate of relative fertility of subjects is grossly in error an unknown factor is determining some of the mutant gene frequencies.

5. That such an inquiry is a big undertaking requiring time, money and trained staff, and that the value of the data collected are dependent on a considerable amount of other knowledge of the community, consanguinity rates, data on fertility, parental ages, age and sex distribution of the population, recording of the frequency of anomalies over a period of time, and so on.

6. That the more highly organised are the medical and social services the better are the prospects for gaining accurate information.

References

CHEESEMAN, E. A. and STEVENSON, A. C. 1955. Records of Deaf Mutism in Northern Ireland. *Brit. J. Prev. & Soc. Med.,* **9,** 196–200.

STEVENSON, A. C. and CHEESEMAN, E. A. 1956. Hereditary Deaf Mutism, with particular reference to Northern Ireland. *Ann. Hum. Genet.,* **20,** 177–231.

NATURAL SELECTION IN MAN: SOME BASIC PROBLEMS

L. S. PENROSE

The problems listed here have received different amounts of attention. There have been innumerable surveys of the frequencies of genes in various human populations, and almost every year has brought its contribution of new genetic biochemical variants to be incorporated in them. But it is true to say, still, that the majority of the world's populations remains unstudied for the majority of its genetic characters, and that the amount of attention to local variation in gene frequency is slight. It has been a very different matter to identify the precise physiological effect of the presence of any particular variant; the glucose-6-phosphate dehydrogenase (Motulsky 1969) and hemoglobin variants are perhaps the best known (Ingram 1963; Antonini 1965; Livingstone 1967; Weatherall 1967), but otherwise knowledge here has advanced little. There have been a few notable studies of the origin of mutations (e.g. Alstrom and Lindelius 1966 on heredo-retinopathia congenitalis in Sweden; Dean 1971 on porphyria in South Africa) but such studies are essentially only possible where there are competent and accurate records of past matings as well as recognition of a particular character. There remains much to do on the stability of gene frequency equilibria, and this whole field of relative advantages of different zygotes and their impact on population evolution still requires much study; it is urgent, for the increasing frequency of deleterious genes and their load upon the health services are fast passing from an academic to a practical and economic problem. Secular trends in quantitative traits have been primarily investigated in stature, menarcheal age, and fertility, though head form in East Europe (Bielicki and Welon 1964) has received some attention. But Penrose's list still remains a valid guide to problems awaiting enquiry.

THOUGH the notion that man had evolved from an ape-like animal by natural selection was accepted with Darwin's ideas nearly a century ago, the possibility of observing natural selection at work in our everyday life was not at first generally appreciated. Although he was greatly interested in human polymorphism, as shown by normal and abnormal hereditary variants, Darwin paid more attention to competition between species or tribes than between individual types. He assumed, in 1871, with Wallace and Galton, that man, by his intelligence, had to a great extent altered the force of natural selection and, at times, reversed its direction. In primitive communities, the more intelligent members would, in the long run, succeed better than the inferior ones and therefore leave more numerous progeny; but, in civilized societies, members of weak bodily and mental constitution are better able to propagate their kind.

The subject of natural selection in man was taken up in a comprehensive way by Karl Pearson in a series of papers beginning in 1894. He tried to estimate the ranges of fertility in different communities and subgroups and to discover how much of this variation was heritable. He discussed the question in its political aspects and concluded that socialism did not necessarily produce degeneration. Later on, in 1912, he became more cautious and expressed the view that only a very thorough eugenic policy could possibly save our race from the evils which must follow from the antagonism between natural selection and medical progress.

In recent years we have become more preoccupied with the effects of selection on characters determined by specific genes. The way was opened by the theoretical work of Fisher, Haldane and Wright on the gene frequency changes in populations under selection. At the present time attention has shifted to more practical problems because of the great advances in serology and chemistry which have enabled so many human genes to be precisely identified.

Our problems are now more definite than formerly; we can ask questions to which we may expect to get factual answers. The basic questions we ask seem to group themselves as follows:

1. *What genes can be recognized in the human population and what are their frequencies?*
2. *What are their precise effects?*
3. *How did the genes get there, that is, when did the mutations take place which brought them into circulation?*
4. *Are these gene frequencies stable? Are the genes maintaining themselves, increasing in frequency or are they diminishing?*
5. *What are the trends of these processes in quantitative traits?*

Only in the last question do we trespass on the ground which was formerly so attractive for human biological speculation.

1. *What genes can be recognized in the human population and what are their frequencies?*

Emphasis was originally directed towards human characters which could be measured with a ruler. Such traits did not segregate and were not suitable for genetical analysis except by quantitative statistical methods. The physical characters, which actually segregated, were rare and, because they were regarded as abnormal, they were excluded from anthropological surveys. Polydactyly and brachydactyly were traits of this kind and their incidences could easily have been ascertained, in populations all over the world, had the effort been made. As it is, we have extensive data on the frequency distributions of body measurements, cephalic indices, hair colours and eye colours, which are common variations but non-segregating.

Investigators tend to use the tools which come to hand conveniently and the characters studied continue to change accordingly from time to time. The transition from physical measurements to blood groups, initiated by the Hirszfelds, is now being succeeded by a shift from serology to chemistry of serum proteins. The development began with the introduction of techniques of chromatography and electrophoresis. Pauling's successful attack on the chemistry of haemoglobins has been followed by further advances by Ingram. The field was again expanded by Smithies with the introduction of starch gel methods and the detection of new serum protein polymorphisms. These movements have been advantageous to anthropology because, by technical advances, we get nearer to the gene and, thus, each new survey gives more accurate information than the last about the genetical compositions of the different human populations studied.

2. *What are their effects?*

It is insufficient merely to count genes, we must also ascertain what effects these genes have on the individuals who carry them. Colour changes are dramatic but not always easy to connect with single gene chemical differences or enzyme deficiencies, except perhaps in albinism. With stature it is even more difficult, except in the extreme cases of dwarfs. Looked at from the opposite angle, hidden chemical differences, as in the blood groups or serum proteins, often do not appear to have much selective significance. The antigenic incompatibilities produced between mother and foetus took a long time to discover but they are among the best established selective effects of polymorphic traits. Other effects of blood group peculiarities on health will take a long time to pin down with equal precision. On the whole we should not expect common genes to show marked associations with diseases.

3. *How did the genes get there, that is, when did the mutations take place which brought them into circulation?*

Tied together with the problems of gene frequency and gene action is the problem of gene origin. We presume that each allele, good or bad, must have originated some time in the past. There is a tendency to think of most genetic variation as having arisen quite recently but this may not be so. A common view is that mutation is going on at all loci fairly regularly at a rate of $1/10^5$ or $1/10^6$ per generation and that this rate is maintained equally in all populations. Against this view, however, there are many points to consider.

Mutation has not been directly observed in any of the typical biochemical traits or any blood group system. Thus it may be much rarer than supposed. The known cases of mutation, such as achondroplasia and retinoblastoma, may be atypical. Then there are curious distributions of genes, very rare in the world population but common in certain isolated populations. Pentosuria and Tay-Sachs disease in Jewish groups are classical examples; the –D– blood group peculiarity; acatalasaemia in Japan, and so on. The suggestion here is that the casual mutations are examples of excessively rare or unique occurrences. With the enormous increase in population numbers through the centuries it is also possible that some more widely spread known mutants, such as that for thalassaemia, may have arisen only once in the distant past.

Moreover, we cannot, in our present state of ignorance, assume that mutants occur regularly at the same locus, even though they are apparently repeated; the work of Benzer on bacteriophage makes us hesitate to assume that, in such cases, it is necessarily exactly the same point on the DNA chain which is altered on each occasion. So that there is no guarantee that a gene identified in one part of the world for a blood group, an anaemia or a malformation is exactly the same as another. How this consideration will affect anthropology remains yet to be seen.

4. *Are these gene frequencies stable? Are the genes maintaining themselves, increasing in frequency or are they diminishing?*

The question of evolution, which depends upon instability of gene frequencies, ultimately is reduced, in formal genetics, to the action of selection upon genotypes. Most of the work of Fisher, Wright and Haldane was concerned with deducing formal laws about rate of entry and extinction of genes which had good or bad effects in theoretical populations. Changes were expected to be very slow, on the whole, and "fixation" of genotypes in the form of favourable homozygosis was the ultimate destination of selection at each locus. Although the possibility of stable equilibrium, which would keep common allelic genes in circulation permanently, was discussed, it was not considered likely to be very significant, perhaps because it had no obvious bearing on evolution, for changes are slowed down by stable polymorphism. It would be well here just to mention the different kinds of genic equilibrium which may be found.

a. Neutral equilibrium, this occurs when there is no selective advantage for any genotype. With random mating this gives the Hardy-Weinberg rule.

b. Stable equilibrium. There are two distinct types. In the first type, (a) selection is balanced by mutation. This we find, presumably, in such a condition as achondroplasia where infertility is very marked and nearly all cases arise by fresh mutation. Increase of fitness or of mutation would raise the level of incidence though this must remain very low. The disease will not ordinarily be considered as a case of polymorphism because it is so rare. Secondly, there is (b) the important type of stable equilibrium, produced if homozygotes are relatively less favoured than the heterozygotes in an allelic system. The main significance of such a situation is that, by this means, apparently unfavourable genes can be maintained in the population. In the homozygote, the gene can be bad or lethal but, in the heterozygote, it can be good and a balance is struck at the appropriate gene frequency. The effects of mutation, on such a system, only make a slight adjustment to the point of equilibrium. Some principles are worth emphasizing here:

(i) Only a very slight heterozygous advantage is required to balance even a lethal homozygote; in fact it needs to have a proportional advantage equal to the gene frequency. One per cent increase in fertility of carriers of phenylketonuria will more than balance the gene's selective disadvantage in homozygotes. In the main, it will be very difficult to detect the advantage to a heterozygote which keeps a given polymorphism in equilibrium in the human population. However, this should be possible in the exceptional case of a common lethal situation, such as that produced by the sickle cell haemoglobin, where the gene is supposed to be lethal in the homozygote.

(ii) It is useless to expect the other types of selective advantage to act as stabilizers unless we are prepared to specify rather exceptional circumstances. For instance, family replacement is a favourite idea. Suppose that, for every recessive idiot, the parents have another, extra, child, this is not enough to preserve the balance: instead they must have four more children. Again blood group A may be a part cause of gastric cancer, if so it must be only the AA homozygote which is damaging, balanced, presumably, by the OO homozygote, who is subject to ulcers. This would help to keep antigen A stable in the population for then the AO heterozygote would have an advantage. The effect, even if proved, would be extremely slight because of the late ages of onset associated with these diseases. Another example is colour blindness. For stable equilibrium, a hypothetical advantage for colour blind men could be balanced by a disadvantage for colour blind women.

c. Unstable equilibrium. Some situations are known which produce unstable states and even one which produces unstable equilibrium. Disadvantage of the heterozygote produces instability and this is known to occur in the case of maternal-foetal incompatibility. The foetus which is heterozygous is at risk. In 1943, Haldane pointed out that, at the gene frequency of 50%, when the homozygotes were equally good and the heterozygote was bad, there was equilibrium but of an unstable type. In this region, even with the frequency of 40%, like the Rhesus gene *d,* changes would be very slow. A rare immunizing antigen, unbalanced by mutation or heterozygous advantage of some kind, would be very rapidly selected against, like an ordinary bad dominant trait, and would disappear in relatively few generations. Correspondingly, a very common antigen would tend to become universal. Perhaps there has been selection of this kind against antigen B, in favour of O, while A, nearer to 50%, has held its own throughout the world.

The effect of natural selection, besides being influenced by the relative fitnesses and frequencies of the genotypes concerned, is also altered by mating systems. On the whole, inbreeding and assortive mating tend to speed up selection because they boost the numbers of homozygotes. They also tend to prevent stable equilibrium and, in their presence, higher degrees of heterozygous advantage are required to produce stability. Thus, in human populations it is unwise to assume stability until all the factors have been looked into and, if possible, measured.

Some situations, which are difficult to understand, arise in connection with common lethal or semi-lethal traits believed to be single gene homozygotes; for example, cystic fibrosis of the pancreas, with an incidence 1/1000, is typical and perhaps also some kinds of anencephaly. How could the genes for such diseases have reached a high level like 1/30 unless there was a fabulous mutation rate? Personally, I think the answer is that the heterozygotes, at one time or in one climate, in famine or in pestilence, were at a huge advantage but that the genes are now on their way out, moving too slowly to be noticeable.

5. *What are the trends of these processes in quantitative traits?*

With quantitative traits, the situation is not unpromising because here, although the genic background eludes us still, trends can be easily measured and observed. There is a high degree of polymorphism noticeable in the metrically simple but genetically complex traits, like stature, weight at birth and intelligence, even allowing plenty of play for variations due to environmental factors. What preserves this variation, in so far as it is genetical? We naturally look for evidence of stable equilibrium. One possibility is mutation of genes causing both tall stature and short stature, for example, and the balancing of opposing trends. Another and much more powerful influence could be the practical disadvantages of too great height or too small stature. Since, in a metrical trait, the extremes of the scale are mainly produced by homozygotes, selection against extreme measurements is equivalent to heterozygote advantage.

There is little direct evidence about stature and fitness, in the sense of individual fertility, but the position is clearer in relation to intelligence measurement. Here we know for certain that the very low levels are associated with lethality and that, in practice, the high levels are associated with an easily measurable degree of infertility as compared with medium or slightly lower than normal levels. The result of this is to maintain a fairly stable polymorphism of intelligence variation in human populations. It seems odd that this very striking situation was not appreciated by authorities who have confidently but, so far, incorrectly predicted rapid decline in the intelligence levels of populations in consequence of fertility differentials; any such adverse effect would be most strongly buffered by the tendency to genetical stability. In general, the multiplicity of polymorphic systems which exists in man probably has an evolutionary value in the long run. Fisher once aptly called the variance in a genetical trait the "energy" of the species and polymorphism, in this sense, stores energy to combat future environmental changes.

The belief that natural selection in man has been abolished by civilization, socialism, hygiene or whatever it may be, depends upon superficial reasoning. What has happened is that the force has been altered and transferred at certain points from one genotype to another. Going back to Darwin (1871), the lower animals must have their bodily structures modified in order to survive under changed conditions. "They must be rendered stronger or acquire effective teeth or claws in order to defend themselves from new enemies; or they must be reduced in size so as to escape detection and danger", whereas man invents weapons, tools and various stratagems, by means of which he procures food and defends himself. In doing this he has reduced the selective values of one set of genes and increased the advantages of others. This change is clearly seen with genes whose reason for existence was to fight infectious diseases now controlled by other means. The genes no longer give this advantage. For example, the thalassaemia trait will probably slowly diminish in frequency now that there is little malaria. Cure of homozygous Cooley's anaemia would, however, block this decline.

The cure of rare cases of known dominant hereditary diseases will not have much effect on the population because the genes concerned are rare usually and slight increases in the prevalence of curable hereditary diseases will not be a biological catastrophe.

The main force of natural selection now seems to be directed towards defects present before birth and leading to failure of development or of function. Even in highly civilized countries, like the United Kingdom or the United States, nearly half of all zygotes formed are unfit in the crude sense of failure to reproduce and it may be assumed that this failure is, to a significant degree, attributable to the genes carried by them. By piecing together evidence from many sources, I have estimated that early prenatal loss accounts for at least 15%; then 3% of the remainder are stillborn, 2% are counted as neonatal deaths and 3% more die before reaching maturity. Of the survivors, 20% do not marry, and, of those who do, 10% remain childless. In view of the large extent, and the persistence, of this loss and the rarity of observed mutation, it seems probable that selection is, for the most part, acting on homozygotes at both ends of the scale, keeping the population in genetical equilibrium. There may be some multiple allelic or pseudoallelic systems present with all the homozygotes lethal. For practical purposes such systems are ineradicable and, if the lethal effect is shown at early embryonic stages only, they will not be harmful.

Consideration of the mechanism of balanced polymorphism goes far towards understanding the problem of selection in man under civilized conditions. The processes involved are complex but we can confidently infer that changes in gene frequency are likely to be slow. Physical anthropologists may be reassured that, considered within historic times, the degrees of population mixtures will probably continue to be accurately measured by gene frequencies.

References

DARWIN, C. (1871) *The Descent of Man.* John Murray, London.
FISHER, R. A. (1930) *The Genetical Theory of Natural Selection.* Clarendon Press, Oxford.

HALDANE, J. B. S. (1943) Selection against heterozygosis in man. *Ann. Eugen., Lond.,* **11,** 333.

HIRSZFELD, L. and HIRSZFELD, H. (1919) Serological differences between the blood groups of different races. *Lancet* **2,** 675.

INGRAM, V. M. (1957) Gene mutations in human haemoglobin: the chemical difference between normal and sickle cell haemoglobin. *Nature* **180,** 326.

PAULING, L., ITANO, H. A., SINGER, S. J. and WELLS, I. C. (1949) Sickle-cell anaemia, a molecular disease. *Science* **110,** 543.

PEARSON, K. (1894) Socialism and natural selection. *Fortnightly Rev.* **56,** 1.

PEARSON, K. (1912) Darwinism, medical progress and eugenics. *Eugenics Laboratory Lecture Series* IX. Cambridge University Press, London.

PENROSE, L. S. (1955) Evidence of heterosis in man. *Proc. Roy. Soc.* B **144,** 203.

SMITHIES, O. and FORD WALKER, N. (1955) Genetic control of some serum proteins in normal humans. *Nature* **176,** 1265.

WRIGHT, S. (1931) Evolution in Mendelian populations. *Genetics* **16,** 97.

MATHEMATICAL MODELS FOR SELECTION

A. R. G. OWEN

In this paper, Dr. Owen drew attention to the potentially complex situations of gene frequency dynamics, neglected in the earlier simple mathematical formulations restricted to a pair of alleles at a single locus. The development of polymorphic studies revealed many loci at which more than two alleles are present, and here Owen argued the existence of a heterozygosity principle of general application. There have been few attempts to examine human populations in the light of this reasoning, and such as there are suggest that few populations can be regarded as being in stable equilibrium. Owen's work has been continued in other complex situations, e.g. Arunachalam and Owen (1971) dealing with polymorphism involving linked loci. At the same time there has been considerable activity questioning the assumptions and applicability of the earlier simple formulations (e.g. Turner 1970; Kimura 1968; Moran 1962) all endeavouring to refine the mathematical models fundamental to understanding the processes of evolutionary genetics.

FOR a long time population geneticists got on very well with only the simplest models of selection acting on genotypes. For instance, if the genotypes available with two alleles A and B have fitnesses in the order AA fittest, AB less fit, BB least fit, then the gene A is unconditionally advantageous and will rapidly replace B in any sizable population. This is exemplified by the spread of melanic genes in European Lepidoptera.

The other simple model (Fisher, 1922) which has been immensely stimulating to research concerns the balanced polymorphism in which the genotypes AA, AB, BB maintain constant proportions. Such a polymorphism exists only when the heterozygote AB is fitter than either homozygote. This situation appears to be realized by the haemoglobin genes A and S in regions of Africa.

Mathematical geneticists have tended to neglect the more complicated cases. For example, the situation with two alleles segregating at each of two loci is also unexplored. Yet even this problem is simpler than many of great interest at the present time, such as the interaction of the ABO and Rhesus systems. The reason for this lag in theory may be merely that at first sight the mathematics is unappetizing and promises little in the way of general principles of a graphic nature of the kind that the practical scientist wants as conceptual tools. However, the experience of myself and my colleague Mr. S. P. H. Mandel with one particular problem (Owen, 1954; Mandel, 1959a) has shown that this pessimistic outlook may be quite illusory. My present belief is that most of these problems admit of conceptually attractive solutions awaiting discovery by a sufficiently pertinacious attack.

The problem we have studied is that of the polymorphism of a series of more than two alleles. The principles involved may be adequately illustrated by the case of three alleles A, B, C. We find that we have to rearrange our concepts concerning the mechanism of a balanced polymorphism. The case of two alleles is covered by the simple "heterozygosity principle"; "A polymorphism results if and only if the heterozygotes are fitter than the homozygotes". With three alleles the principle in this form just does not apply. For instance, if all homozygotes are lethal, a polymorphism only results if the fitnesses f, g and h of the heterozygotes BC, AC and AB satisfy the same quantitative relation as do the lengths of the sides of a triangle, namely

$$f < g + h, \quad g < h + f, \quad h < f + g.$$

For example, with lethal homozygotes and relative fitnesses 10, 5, 4 no polymorphism results; at least one gene is completely eleminated by selection. (I should mention, perhaps, that no biological significance attaches to the triangle whose presence is due only to mathematical coincidence.)

Contrariwise, we can have polymorphisms (with A, B and C all represented) in cases when some of the homozygotes are fitter than some of the heterozygotes. For example, when the genotypic fitnesses are:

AA	BB	CC	BC	CA	AB
0·6250	0	0·3125	1·7500	1·5000	0·500

then the homozygote AA is fitter than the heterozygote AB. However, a stable polymorphism exists with gene frequencies A 41·44%, B 11·11%, C 44·44%. The stability index (a measure of the degree of stability of the polymorphism) is 0·08492.

A more extreme case has *two* homozygotes fitter than one heterozygote. The genotype fitnesses are:

AA	BB	CC	BC	CA	AB
0·7188	0	0·8848	0·6875	1·1797	1·500

and the polymorphism has gene frequencies A 41·38%, B 3·45%, C 55·171%, with stability index 0·00391.

It will be noted that the stability index diminishes with decreasing overall superiority of the heterozygotes, and this suggests that a heterozygosity principle is still operating. This conclusion is supported by the demonstrable facts that in a stable polymorphism of three alleles no heterozygote (such as AB) can be less fit than *both* of its associated homozygotes (AA, BB), and not more than one heterozygote can be less fit than a homozygote. Furthermore, in the general case of any number of segregating alleles it has been shown that for a polymorphism to exist the fitness of each homozygote must be less than the average fitness of the population (Mandel 1959a).

These results all indicate that a heterozygosity principle is functioning. One formulation of the new heterozygosity principle can be illustrated by a reconsideration of the balanced polymorphism of two alleles A and B. All geneticists are familiar with Wright's inbreeding coefficient F. This coefficient is a measure of the degree to which the heterozygotes are deficient in number in comparison with the case of random mating without selection. As we wish to be free of the connotation of inbreeding, being concerned with random mating, we use a coefficient φ equal to $(-F)$. If we use the symbol for each genotype (such as AB) to denote also the proportion of the total population constituted by individuals of that genotype then $\varphi_{AB} = [(\frac{1}{2}AB)^2 - (AA)(BB)]/4p_A p_B$ where p_A and p_B are the gene frequencies of A and B. Then the principle of heterozygosity can be reformulated in mathematical terms as "$\varphi > 0$". Thus in the two allele case the condition for a polymorphism to be stable is that it be characterized by an excess of heterozygotes in comparison with what would be attained (with the same gene frequencies) under random mating without selection (i.e. in comparison with Hardy-Weinberg expectations for the same gene frequencies).

Exact algebraical treatment shows that a polymorphism is stable if, and only if, all the possible φ coefficients which can be defined are positive. Thus with three alleles using a determinant notation, necessary and sufficient conditions for stability of a polymorphism are two in number, e.g.

$$\varphi_{AB} = - \begin{vmatrix} AA & \frac{1}{2}AB \\ \frac{1}{2}AB & BB \end{vmatrix} > 0,$$

$$\text{and } \varphi_{ABC} = \begin{vmatrix} AA & \frac{1}{2}AB & \frac{1}{2}AC \\ \frac{1}{2}AB & BB & \frac{1}{2}BC \\ \frac{1}{2}AC & \frac{1}{2}BC & CC \end{vmatrix} > 0$$

When these conditions are both satisfied then it follows automatically that φ_{BC} and φ_{AC} are both positive also. The requirements $\varphi_{AB} > 0$, $\varphi_{BC} > 0$, $\varphi_{AC} > 0$ indicate that every heterozygote is present relative to its associated homozygotes in excess of Hardy-Weinberg expectations. The requirement $\varphi_{ABC} > 0$ is a somewhat more subtle manifestation of the principle of heterozygosity. The connection between this formulation and the necessary conditions mentioned above requiring each homozygote to be less fit than the population average is still in process of being worked out.

Success with the multiple allele problem encourages me to think that more complex problems will be reducible to relatively simple heterozygosity principles. Numerical work by Parsons (1959) on populations of tetraploids supports this conclusion and Mandel (1959b) has in the case of sex-linked genes rescued the heterozygosity principle from the discredit into which it temporarily fell in consequence of a recent paper of Bennett (1957). Rehabilitation may also be possible in the case of sex-limited autosomal genes (Owen, 1952).

In this discussion I have considered only selection in its simplest form, as acting solely on genotypes. But the Rhesus situation exemplifies the more general possibility in which selection acts differentially on like genotypes according to their parentage. Even when a small number of alleles is involved (such as two only), we have to envisage a wealth of possibilities. Oscillatory systems, for instance, cannot be ruled out. Mr. Mandel and myself have entertained ourselves with models of systems which are periodic, moving through a closed cycle of different polymorphic states. Again a more complex model (Owen, 1953) with three alleles shows complete instability, the population never coming to equilibrium. Our models are admittedly *jeux d'esprit,* but we submit that there is a great deal to be thought about in this field.

One last question that ought to be elucidated is the validity of approximations based on the assumption of infinitesimally slow selection. These approximations are usually mathematically tractable, which I take to be one of the cardinal reasons for their popularity, and it is indeed a highly respectable reason. There are occasional disasters where the assumption leads to results which are not merely quantitatively but qualitatively in error. Such acute failures are no doubt restricted to situations where the polymorphism is unstable. Nevertheless, the question of validity deserves investigation. Often the assumption of very slow selection is supposedly justified by the argument that in the populations concerned generations overlap. But this is to say that because a process is continuous it must be slow. I am unhappy about this thesis and trust that it will be given critical study.

References

BENNETT, J. H. (1957) Selectivity balanced polymorphism at a sex-linked locus. *Nature* **180,** 1363.

FISHER, R. A. (1922) On the dominance ratio. *Proc. Roy. Soc. Edin.* **42,** 321.

MANDEL, S. P. H. (1959a) The stability of a multiple allelic system. *Heredity* **13,** 289.

MANDEL, S. P. H. (1959b) Stable equilibrium at a sex-linked locus. *Nature* **183,** 1347.

OWEN, A. R. G. (1952) A genetical system admitting of two stable equilibria. *Nature* **170,** 1127.

OWEN, A. R. G. (1953) A genetical system admitting to two distinct stable equilibria under natural selection. *Heredity* **7,** 97.

OWEN, A. R. G. (1953) Genetical Society Abstracts. Equilibrium of populations and the possibility of sustained large-scale oscillations. *Heredity* **7,** 151.

OWEN, A. R. G. (1954) Balanced polymorphism of a multiple allelic series. *Proc. 9th Int. Congr. Genetics, Caryologia,* Suppl. to vol. 6, p. 1240.

PARSONS, P. A. (1959) Equilibria in autotetraploids under natural selection for a simplified model of viabilities. *Biometrics* **15,** 20.

THE RELATIVE FITNESS OF HUMAN MUTANT GENOTYPES

By C. A. CLARKE

It is surprising how little attention has been given to the biological fitness of genotypes producing clinical conditions, for fitness estimates form the basis of predictions of the response of gene frequency to changed patterns of treatment about which there has recently been so much discussion (e.g. Motulsky *et al.*, 1971; Fraser, 1972). This paper remains among the most comprehensive reviews of the subject. With the improved biochemical methods of identifying carriers of genes (both for recessive diseases and for dominants of variable expression) and of distinguishing genetically distinct subdivisions of clinical entities, there is great scope for the future development of this type of work. However, an interesting development has been the recognition of the importance of the variations that clinical phenotypes confer in social, as distinct from biological, fitness in human society (Walker *et al.*, 1971; Boon and Roberts 1970).

THE fitness which is being discussed in this paper is Darwinian, biological or reproductive fitness, and in respect of a mutant gene it is the extent to which an individual possessing that gene can reproduce so that the gene is maintained in the population. Fitness is thus a measurement of net fertility, that is, the number of offspring who reach the mean age at which the parent reproduced (see Fisher, 1930; Penrose, 1949, 1950). It is not a synonym for fertility since obviously a large number of offspring dying before maturity do not contribute at all to the next generation. The time factor is also important, biological fitness implying the capacity of a trait to be maintained through many generations in differing environments (see Thoday, 1953).

Fitness is said to be unity if the individual produces one child which survives to the age of the parent, but in practice "children require two parents to produce them and, therefore, two parents, each with unit fitness, have to have two children to maintain the equilibrium of any given gene in the population" (Penrose, 1949).

THE MEASUREMENT OF FITNESS

(a) *Dominant Genes*

The most comprehensive method of estimating the fitness of a dominant trait is that of direct ascertainment, whereby all individuals with the trait in a given area are investigated and their reproductive performance compared with a control series. The classical example is that of Mørch (1941), who traced all the achondroplasic dwarfs in Denmark. In the population of $3\frac{1}{4}$ million there were 108 achondroplasics and Mørch found that they had produced between them only 27 children, whereas their 457 unaffected sibs had produced 582 offspring. Clearly, the fitness of the dwarfs was very low since only one in four of them on an average had produced offspring, while their unaffected sibs had produced about $1\frac{1}{4}$ children each. From Mørch's data the exact figure for the fitness (f) of the abnormal allele can be calculated as 0·098 and thus 90·2% of the chondrodystrophic genes are eliminated in one generation.

Mørch also noted that of his 108 dwarfs only 18 gave a positive family history of the condition. As it is known that achondroplasia is controlled by a dominant gene of high penetrance, Mørch assumed that the majority of his cases were due to mutation, and he calculated that the mutation rate for achondroplasia was 48 per million loci per generation (the "direct method"). An estimate of the mutation rate can also be made from a knowledge of the fitness and the incidence of the trait (the "indirect method"), and conversely, if the mutation rate and the incidence are known the fitness of the gene can be deduced. When this method is used it is assumed that the population is in equilibrium—that is, the frequency of the trait concerned is not altering. The formula for calculating the relative fitness by the indirect method is as follows (Stern, 1950):

$$f = \frac{1 - m}{q}$$

where f is the relative fertility of individuals heterozygous for the particular gene under consideration, m is the mutation rate per gene per generation, and q is the frequency of the gene—i.e., half the fraction given by the ratio of all cases born during a given period to all births during the same period.

Where a dominant character has a high fitness complete ascertainment is extremely difficult, and it may be impossible to trace all cases in a given area since many of those affected have little disability. Nevertheless, a relative fitness of as high as 0.9 will in ten generations lead to a decline to $(9/10)^{10}$ or to less than 4% of the original frequency. Consequently, complete ascertainment is highly desirable, the examination of pedigrees being rather uninformative unless the trait being investigated is unfavourable.

(b) *Recessive Genes*

The important point here is that genes controlling recessive traits are much more numerous in a population in the heterozygous than the homozygous state. However, since the carriers of disease can rarely be identified with certainty, it is impossible to calculate accurately the fitness of recessive genes. Estimates based on the assumption that genes when heterozygous are neutral are fallacious, since from other evidence it is known that heterozygotes often possess some selective advantage or disadvantage (Fisher, 1930). With modern techniques, however, it is becoming possible to identify the heterozygote with increasing accuracy in a number of recessive conditions, and more information about the fitness of carriers—and, therefore, of recessive genes—is likely to be forthcoming.

(c) *Sex-linked Recessive Genes*

The situation here is more favourable than with recessives since the effects of the gene will be demonstrable in 50% of the cases, i.e., in hemizygous males. Since, however, it is not usually possible to identify female carriers, the same objections hold for these as for recessive traits.

THE IMPORTANCE OF THE ESTIMATION OF FITNESS

Mutation Rate

An accurate knowledge of the fitness of a given gene is of great help in estimating the mutation rate. For a dominant it can corroborate the figure obtained by the direct method (see page 2), and the same would be true of a recessive if the heterozygote could be identified. Where this is now possible it would be a practical proposition to score individuals heterozygous for a given trait (say in a large hospital) and in a forward survey assess their fitness.

In man heterozygous advantage has been demonstrated in certain circumstances, notably that of sickling in malarious areas (Allison, 1954). It would be of great interest to know for certain whether those individuals heterozygous for sickling (or thalassaemia) are biologically fitter than normal individuals from the same area (see Raper, 1956; Allison, 1956a and b). It may well be that hybrid vigour can endow an individual with considerable ability to resist infections, and this could have had a great bearing on evolution in the past (Haldane, 1949a).

Recognition of a Polymorphism

In some situations it is uncertain whether a disease is due to the existence of a polymorphism with the heterozygote at an advantage, or whether differing forms of the disease are determined by genes at several loci, each with their own mutation rate. Diabetes mellitus is a problem in this respect. The juvenile form, however, is probably controlled by a recessive gene and it would be of great help if some test could be designed to uncover the heterozygotes and thereafter estimate their fitness. Unfortunately, the administration of cortisone followed by a glucose tolerance test probably only detects the homozygotes who have not developed the disease.

DIFFICULTIES IN ASSESSING FITNESS

These are well recognized and can be summarized as follows:

1. Controls

At first sight it would seem that the unaffected sibs of patients would be the correct control. They are of the same social and economic strata, they are roughly contemporaries and their environment is likely to be similar. Moreover, it is extremely difficult to match an affected individual with an exactly comparable control from the general population. On the other hand, it is not possible, using sibs, to pair a patient with a control of the same age (except with non-identical twins), and in deleterious traits, particularly those affecting mental processes, there may be a tendency for the family to sink in the social scale and to marry into other sub-normal families. Such assortative matings may well reduce the fitness of the unaffected sibs (Leese, Pond and Shields, 1952).

2. Method of Ascertainment

The larger and more fertile the family with an inherited abnormality, the more attention it is likely to receive and, therefore, if general population controls are used the fitness of those with the trait may be assessed too highly. However, if the unaffected sibs are used this objection seems invalid. Furthermore, where there are several forms of the disease due to multiple alleles or genes at different loci, a large family will give a better index of fitness than pooled data.

3. Diagnosis

It is becoming increasingly clear that differing grades of severity in many inherited diseases may be the result of control by a multiple allelomorphic system or by genes at different loci. For example, "haemophilia" (which used to be considered a distinct clinical entity), has recently been shown to vary considerably with regard to the amount of anti-haemophilic globulin present in different families, and also to include a group of conditions due to deficiency of other clotting factors, sometimes with differing modes of inheritance. Each of these may have a different mutation rate and value for fitness, and there is clearly a risk of grave error if family data are pooled. Furthermore, it should be noted that the recent subdivisions of haemophilia will mean that the earlier estimations of the mutation rate will have been too high. It is obvious, therefore, that for the accurate estimate of fitness investigators should be studying exactly the same disease.

Another diagnostic problem may occur in achondroplasia. Mørch's data have been extensively used to calculate the mutation rate of the gene. However, it is now known that more than one achondroplasic can occasionally be born to normal parents, and in one of two such families reported by Stevenson (1957a) the parents were full cousins. This strongly suggests that the condition can be controlled by a recessive gene. If this be true, or if, alternatively, incomplete penetrance of a dominant gene is invoked, the figure for the mutation rate would be too high (Stevenson, 1957b).

There is yet another difficulty; Mørch was of the opinion that the stillbirth and neonatal mortality rate in achondroplasia was very high, whereas Stevenson (1957a) thinks that the evidence is in favour of the numerous cases dying in infancy being due either to a different allele or to environmental causes (phenocopies) or to a confusion with other conditions such as Morquio's disease (osteo-chondrodystrophy). If these early deaths are in fact phenocopies, this would explain the deficiency of dwarfs noted by Haldane in family data (Haldane, 1949b).

These considerations emphasize that even classical data need very careful examination in the light of modern views on diagnosis.

4. Incomplete Penetrance

Not uncommonly in dominant traits there is incomplete penetrance or variable expressivity due either to genetic interaction, to modifiers, or to suppressor genes. If in a family one of these individuals is taken as a control, errors in incidence and fitness (and consequently in the mutation rate) will occur.

In achondroplasia it is of interest that after at least 5000 years (two of the Egyptian gods are represented as achodroplasics) the gene responsible still appears to have full penetrance and does not appear to have been modified appreciably by selection.

5. Low Fitness of Fresh Mutations

As Penrose (1936) has pointed out, an individual carrying a fresh mutant is likely to be less fit than one who has inherited the trait, for in the latter selection pressures will have tended to maintain the less severe forms of the disease.

6. Phenocopies

These are environmentally produced abnormalities which mimic traits known to be genetically determined. Thus some examples of retinoblastoma are thought to be due to somatic mutations and the condition is then not transmitted to descendants. These may be scored as possible sporadic cases by the unwary. Furthermore, the early ophthalmological appearance of retinoblastoma may sometimes be simulated by retrolental fibroplasia, a disease resulting from excess oxygen given to premature infants, and if this mistake in diagnosis is made, eyes may be removed in error. Deaf-mutism, indistinguishable from the inherited form, as Stevenson (1957c) has shown, can be produced by maternal rubella in the early months of pregnancy. Obviously, if any of these phenocopies is used to determine fitness, gross inaccuracies will be introduced.

7. Fitness and the Environment

Changes in the environment may radically alter fitness. Thus the prevention of malaria will remove the advantages of the haemoglobin S gene for sickling, rendering it less fit (Penrose, 1955). On the other hand, modern medical and surgical treatment may improve the fitness of deleterious genes—e.g., the surgical treatment of retinoblastoma and the insulin treatment of diabetes mellitus. There is also at least one inherited condition in which medical science helps to get rid of disadvantageous genes by killing off (unwittingly) individuals with the trait. This occurs in porphyria, where anaesthesia by barbiturates such as pentothal may be fatal (Dean, 1955).

8. Examination of Pedigrees

These are often incomplete and from the point of view of fitness are liable to be unhelpful since the marriage rate and fertility of unaffected sibs are not usually mentioned.

EXAMPLES OF FITNESS IN CERTAIN DISEASES

1. Peroneal Muscular Atrophy

This is a condition beginning usually in childhood or young adults in which there is marked wasting and deformity of the legs and feet and later of the hands; its frequency is about 30 per million of the population. It does not tend appreciably to shorten life, but the economic handicap of the patients (particularly the males) is great. The inheritance is by any of the three classical methods, but in the majority of cases the gene controlling it is dominant to its normal allele. Julia Bell (1935) was of the opinion that the disease did not impair fertility, and Table I, which shows a part of four of her pedigrees, demonstrates this. It will also be seen that more of the affected than unaffected appear to be married, but sometimes the relevant data are not available. It does seem, however, as if most of Julia Bell's trait bearers did, in fact,

TABLE 1. Fitness in peroneal muscular atrophy
(dominant form) from Julia Bell's (1935) 4 pedigrees

Parents	Offspring	Average
50 affected	189	3·62
27 unaffected	85	3·14

marry, and it may be that their energy, vigour and courage (on all of which she comments) helped them to find mates. This is not unlikely since selection will favour genes controlling somatic fitness. Furthermore, in assessing reproductive fitness in civilized communities great importance must be placed on psycho-social factors. Thus compassion, boredom, mutual poverty and proximity all play important parts, but varying in different economic situations. It seems certain that peroneal muscular atrophy would have been highly disadvantageous in primitive man, and that selective pressures have relaxed in modern times.

2. Albinism

In this condition, which is usually inherited as a recessive, there is no means of identifying the heterozygote and, therefore, it is impossible to say anything precise about the fitness of the gene. In the homozygote there is no doubt that it carries many disadvantages, albinos being particularly liable to infections and sensitive to strong sunlight. There may have been less selection against the gene among cave dwellers, and the albino Noah (Sorsby, 1958) will have enjoyed, at any rate, temporary relief from his affliction. In some primitive peoples albinos are avoided; thus in South America among the Indians of the San Blas coast, marriages between albinos are prohibited by the tribe and "white" women rarely marry brown men. They have, however, persisted in numbers among these primitive people for at least three centuries (Gates, 1948). There is, therefore, at least a case for thinking that the heterozygote may be at an advantage.

3. Fibrocystic Disease of the Pancreas

Here there is stunting of growth, diarrhoea and a liability to chest infections due to cystic changes in the pancreas, lungs and other parts of the body; in addition, the sweat contains a higher concentration of sodium chloride than normal. The disease is controlled by a recessive gene. The fitness of the homozygote has usually been stated to be very low, practically no cases surviving to puberty. However, the picture is changing with modern treatments, particularly on account of antibiotics which control the chest infection and pancreatic extract which relieves the diarrhoea. Anderson (1958), who has recorded 500 cases of the disease since 1935, has stated that 275 of them are still alive and 85 of that number are in the second decade of life. A few of these have married. The fitness of the phenotype will not only be influenced by medical measures but by climatic conditions. In a hot country where sweating is profuse severe salt loss may occur and be a cause of death.

The heterozygote is not scorable in any clear-cut manner, though there is a suggestion that in carriers the sodium chloride content of the sweat is often intermediate between normal and affected. Since the frequency of the disease is fairly high (Stevenson in Northern Ireland gives a figure of 600 per million of the population), and the fitness rather low, a high mutation rate has been assumed for the condition, $0.7 - 1.0 \times 10^{-3}$ (Goodman and Reed, 1952, quoted by Neel and Schull, 1954). This, by analogy with other genes is likely to be far too high a figure and it would, therefore, be of particular interest to investigate the fitness of the heterozygote.

4. *The Pelger-Huët Anomaly*

This is an inherited condition controlled by a dominant gene, in which the polymorphonuclear leucocytes are abnormal, having fewer lobes than usual. The trait has also been described in rabbits (Nachtsheim, 1950), where it has been found that homozygous Pelger produces severe skeletal deformities. With one possible exception, that of an epileptic resulting from a cousin marriage between two Pelger individuals (Klein *et al.*, 1955), the homozygous state has never been described, but Nachtsheim has calculated that the gene has a selective disadvantage of 20% in man.

TABLE 2.   Fitness in Pelger Anomaly, 3 Generations
(dominant form) from Davidson, Lawler and
Ackerley's (1954) Pedigree

Parents	Offspring	Average
34 affected	34	1·00
51 unaffected	30	0·58

This seems *a priori* to be unlikely, and in two British families with the condition (Davidson *et al*, 1954; Finn, 1958), those affected have as many offspring as the normals. Table 2 indicates the fertility of part of Dr. Davidson's family.

A possible disadvantage of the Pelger-Huët phenomenon might be the inability to produce a satisfactory polymorphonuclear response in acute infections. In practice, however, those with the anomaly appear to withstand such diseases as well as normal people.

5. *Huntington's Chorea*

This disease, which is characterized by choreic movements of the limbs and later by mental changes, usually does not begin until about the age of 35 years, and it progresses to insanity in middle life. The gene controlling it is dominant to its normal allele. One of the early symptoms is said to be increased libido and this may be a factor in maintaining the gene in the population.

Table 3, taken from Julia Bell's material (1934), shows that the fitness of the heterozygote relative to the unaffected sibs is above unity. This is in agreement with a survey carried out in Michigan, where the affected individuals were found to have produced an average of 6.07 ± 0.9 children, while their unaffected sibs had an average of 3.33 ± 0.5 (Reed and Palm, 1951). In a more recent

TABLE 3.   Fitness in Huntington's Chorea from
Julia Bell's (1934) Pedigree

Parents	Offspring	Average
12 affected	55 (32 affected)	4·58
11 unaffected	37 (none affected)	3·36

investigation choreics were found to have a fitness of 1·03 when a conventional sib comparison was made (Reed, 1958). However, when normal Michigan families were used as controls the fitness of the Huntington heterozygotes fell from 1·03 to 0·79, and Reed is of the opinion that unaffected sibs are not good controls. One reason for this has been referred to earlier (see page 34). Another explanation might be that unaffected sibs would be less likely to marry if they knew that they might carry the taint. However, since the disease does not usually start until the age of 35 the same argument would usually apply to those who were to be affected as to those who were to remain free.

On the assumption that the fitness may be high, it is worth considering whether there are not many undiagnosed cases in asylums—the disease might be confused with presenile dementia, the arteriosclerotic type of Parkinson's disease and cerebral arteriosclerosis (Pleydell, 1954).

6. Retinitis Pigmentosa

This eye complaint, which generally starts in childhood, has as an early symptom inability to see in the dark and it usually terminates in complete blindness. It must have had very grave disadvantages in primitive communities but in modern England those affected seem to fare quite well. This is shown by a large family known to us in Liverpool, in which the disease is inherited as a dominant. Taking that part of the pedigree which is complete (Table **IV**) it will be seen that 7 affected individuals have had 28 children, whereas the 4 normals have only produced 9. One sibship is, of course, very little evidence, but looking at the whole pedigree there is the same suggestion that the fertility of those affected is

TABLE 4.   Fitness in Retinitis Pigmentosa (dominant form)

Parents	Offspring	Average
7 affected	28	4·00
4 unaffected	9	2·25

about normal and, moreover, most of them are married. This again emphasises the importance of psycho-social factors, for it seems probable that marriage gives them some consolation for their tragedy—and waiting in out-patients is a fruitful source of romance, often with other blind or partially-sighted people.

While on the subject of visual defects it will be recalled that Penrose has suggested that in red-green colour blindness the "uncovering" of cryptic coloration might have been advantageous to primitive hunting man, and in modern warfare the colour blind may sometimes be of use in picking out camouflaged targets. In a more peaceful sphere, Ford (1955) has shown that a colour blind entomologist has no difficulty in collecting caterpillars of the emperor moth which, to normal people, merge completely with the background of the heather.

7. Haemophilia

The fitness of the gene responsible for this disease has been estimated in Denmark as 0·286, that is, only 28·6% of the abnormal alleles are transmitted from one generation to the next, 71·4% being eliminated (Stern, 1950). This estimation, however, assumes that the gene is neutral in the heterozygous state.

We thought it of interest to investigate the pedigree of the Royal Family because (a) it is easy to get information about the offspring of both those affected and unaffected, (b) in the different generations the severity of the disease is likely to be fairly uniform, and (c) there will be a comparable knowledge within the family as to the likelihood of passing on the disease.

Table 5 shows that seven known carrier females (including Queen Victoria) produced 37 children, whereas the 15 unaffected females produced only 53. In the affected males the situation was entirely different, for here only one out of eleven produced any offspring. Because of the very poor fitness of the males (for obvious reasons) the frequency of the gene will diminish in succeeding

TABLE 5.   Fitness in Royal Family (Haemophilia) 4 Generations

| Parents | | Affected Sibs | | Unaffected Sibs | |
		1 affected male other 10 unmarried	7 carrier ♀♀	9♂♂	6♀♀
Offspring	Affected ♂♂		11		
	Carrier ♀♀	1	5		
	Unaffected ♂♂	1	8	11	13
	Unaffected ♀♀		13	14	15
	Totals	2	37	25	28
		39		53	

generations, but the carrier females do not seem to be at a disadvantage in this family. Moreover, a study of the pedigree does not suggest that the phenomenon of "compensation" is operating.

8. *Tylosis*

This is an inherited condition characterized by hyperkeratosis of the palms and soles (see Fig. 1). It is controlled by a dominant gene with high penetrance, and is generally considered to be a rather rare and trivial disability (incidence 35 per million, fitness 0·95, Stevenson, 1958).

We have investigated two families (possibly related), and whereas 27 of those heterozygous for the trait produced 99 offspring, their 26 unaffected sibs had 83 children. Taken at their face value, these findings do not suggest that there is any reproductive disadvantage

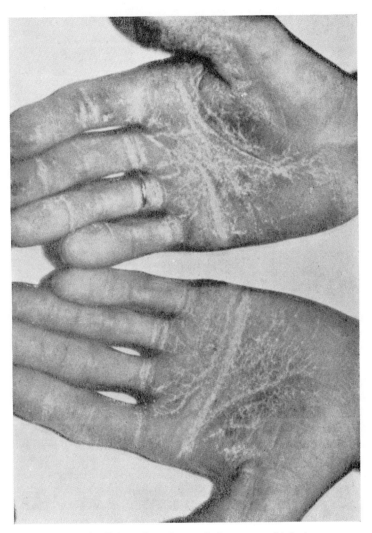

FIG. 1.   Palmar hyperkeratosis in a case of tylosis.

in being tylotic, and it is of particular interest to note that in our families there was also a marked association between the skin lesion and carcinoma of the oesphagus, 18 patients being affected

TABLE 6. Fitness in Tylosis (two pooled families in which the skin condition was associated with carcinoma of oesophagus

Parents	Offspring	Average
Tylotics		
14 without carcinoma	59	4·21
13 with carcinoma	40	3·07
Non-tylotics		
26	83	3·19

by the growth (Howel-Evans *et al.*, 1958). Thirteen of these were among the 27 parents mentioned above, and they produced 40 of the 99 offspring (see Table 6). Although the growth developed at a comparatively early age, usually between 30 and 45 years, it did not seriously lower their reproductive performance.

Discussion

On looking through text books on human genetics one is struck by the number of rare abnormalities controlled by dominant genes, and moreover, in many of these the disability appears to be rather trivial. The recessive traits, on the other hand, are less numerous but on the whole more deleterious. The reasons for this are: (i) dominant traits are much more readily recognizable when heterozygous for the gene than are autosomal recessives, and the frequency of sex-linked recessive traits in man can be explained by the fact that the hemizygous males are often easily identifiable. (ii) The high fitness of many of the dominant mutants will mean that the Fisher effect, whereby mutants which are originally intermediate become recessive if deleterious, is only slightly operative (Fisher, 1930; Levit, 1936). (iii) Some of the genes which we describe as "dominants" would be called recessives if they occurred more often in their much more lethal homozygous form (Haldane, 1939). (iv) The rarity of lethal and sublethal recessive conditions is partly accounted for by the relaxation of inbreeding which has occurred during historical times in most civilized communities (Haldane, 1939).

Although the data in our pedigrees are probably biased because of the inclusion of large fertile families which have attracted special attention, yet there seems no doubt that many dominant mutant genes have, anyhow at the present time, a fitness value of near unity. The problem is to explain why they remain so rare, and the explanation must be that their mutation rates are extremely low, some of the traits concerned possibly being derived entirely from a single individual. The inference is that there has been a general tendency to estimate mutation rates at too high a level. On the other hand, it must be borne in mind that loci vary greatly in their sensitivity to mutagens, and it may be that those which mutate frequently produce the most harmful mutants.

To summarize, it is clear that the investigation of fitness is beset with pitfalls. The subject, however, is one of great interest and with improved techniques new light may be shed on the matter, particularly on the problem of the fitness of recessive genes by study of the heterozygous state. Public interest would be well served if the registration of abnormal traits were to be made compulsory.

Acknowledgements

I am greatly indebted to Professor L. S. Penrose, F.R.S., Dr. G. A. Harrison, Dr. R. B. McConnell and Dr. P. M. Sheppard for advice on the preparation of this paper, and to Dr. A. C. Stevenson for giving me his most informative report on the congenital and hereditary anomalies in the population of Northern Ireland.

References

ALLISON, A. C. (1954) Protection afforded by the sickle cell trait against subtertian malarial infection. *Brit. Med. J.* **1**, 290.

ALLISON, A. C. (1956a) The sickle-cell and haemoglobin C genes in some African populations. *Ann. Hu. Genet.* **21**, 67.

ALLISON, A. C. (1956b) Population genetics of abnormal human haemoglobins. *Proc. Int. Congr. Hum. Genet.* **1**, 430. Copenhagen.

ANDERSON, D. (1958) Address to a clinical meeting at Queen Elizabeth Hospital for Children, London.

BELL, JULIA (1934) Huntington's Chorea. *Treasury of Human Inheritance,* vol. **4**, part 1.

BELL, JULIA (1935) On the peroneal type of progressive muscular atrophy. *Treasury of Human Inheritance,* vol. **4**, part 2.

DAVIDSON, W. M., LAWLER, SYLVIA D. and ACKERLY, A. G. (1954) The Pelger-Huët anomaly, Investigation of family "A". *Ann. Hum. Genet.* **19**, 1.

DEAN, G. and BARNES, H. D. (1955) The Inheritance of Porphyria. *Brit. Med. J.* **2**, 89.

FINN, R. (1958) Personal communication.

48   HUMAN VARIATION AND NATURAL SELECTION

FISHER, R. A. (1930) *The Genetical Theory of Natural Selection.* Clarendon Press, Oxford.
FORD, E. B. (1955) *Moths.* Collins, London.
GATES, R. R. (1948) *Human Genetics* 1, 285. Macmillan, New York. York.
GOODMAN, H. O. and REED, S. C. (1952) Heredity of fibrosis of the pancreas; Possible mutation rate of the gene. *Am. J. Hum. Genet.* 4, 59.
HALDANE, J. B. S. (1939) The spread of harmful autosomal recessive genes in human populations. *Ann. Eugen. Lond.* 9, 233.
HALDANE, J. B. S. (1949a) Disease in Evolution. *Ric. Sci.* 19, (Suppl.).
HALDANE, J. B. S. (1949b) The rate of mutation of human genes. *Proc. 8th Int. Congr. Genetics* (Suppl. to *Hereditas*), pp. 267–273.
HOWEL-EVANS, W., MCCONNELL, R. B., CLARKE, C. A. and SHEPPARD, P. M. (1958) Carcinoma of the oesophagus with keratosis palmaris et plantaris (Tylosis). *Quart. J. Med.* 17 (No. 107), 413.
KLEIN, A., HUSSAR, A. E. and BERNSTEIN, S. (1955) Pelger-Huët anomaly of the leucocytes. *New Eng. J. Med.* 253, 1057.
LEVIT, S. G. (1936) The problem of dominance in man. *J. Genet.* 33, 410.
LEESE, STEPHANIE M., POND, D. A. and SHIELDS, J. (1952) A pedigree of Huntington's chorea, with a note on linkage by R. R. Race. *Ann. Eugen. Lond.* 17, 92.
MØRCH, E. T. (1941) *Chondrodystrophic Dwarfs in Denmark.* Ejnar Munksgaard, Copenhagen.
NACHTSHEIM, H. (1950) Pelger anomaly in man and rabbit; Mendelian character of nuclei of the leucocytes. *J. Hered.* 41, 131.
NEEL, J. V. and SCHULL, W. J. (1954) *Human Heredity.* University of Chicago Press.
PENROSE, L. S. (1936) Autosomal mutation and modification in man with special reference to mental defect. *Ann. Eugen. Lond.* 7, 1–16.
PENROSE, L. S. (1949) The meaning of fitness in human populations. *Ann. Eugen. Lond.* 14, pt. 4, 301.
PENROSE, L. S. (1950) Propagation of the unfit. *Lancet* 2, 425.
PENROSE, L. S. (1955) Evidence of heterosis in man. *Proc. Roy. Soc.* 144, 203.
PLEYDELL, M. J. (1954) Huntington's chorea in Northamptonshire. *Brit. Med. J.* 2, 1121.
RAPER, A. B. (1956) Sickling in relation to morbidity from malaria and other diseases. *Brit. Med. J.* 1, 965.
REED, S. C. and PALM (1951) Social fitness versus reproductive fitness. *Science* 113, 294.
REED, T. E. (1958) Methods and problems in estimating relative fitness in man; Examples from a study of Huntington's chorea. *Proc. Int. Congr. Genetics, Montreal.*
SORSBY, A. (1958) Noah—an Albino. *Brit. Med. J.* 2, 1587.
STERN, C. (1950) *Principles of Human Genetics.* Freeman, San Francisco, Calif.
STEVENSON, A. C. (1957a) Achondroplasia; an account of the condition in Northern Ireland. *Amer. J. Hum. Genet.* 9, 81.
STEVENSON, A. C. (1957b) *Some Data, Estimates and Reflections on Congenital and Hereditary Anomalies in the Population of Northern Ireland.* Dept. of Social and Preventive Medicine, Queens University, Belfast.
STEVENSON, A. C. (1958) Biological studies of small communities. Symposia of the Society for the Study of Human Biology, 1, 11.
THODAY, J. M. (1953) Components of fitness. *S.E.B. Symposia, Evolution* 7, 96.

NATURAL SELECTION AND SOME POLYMORPHIC CHARACTERS IN MAN

P. M. SHEPPARD

In the call here made for a systematic attack on the problems of human polymorphisms, Professor Sheppard singled out four approaches that seemed to be worthy of special attention, and in so doing he was drawing on the knowledge of polymorphisms occurring in nature so usefully reviewed in his book (Sheppard 1958). Each of these has been further investigated to some degree. The retrospective analyses of fertility and abortion have been the subject of some eighteen studies since 1965 alone (Reed 1974). Prospective studies have been carried out on large numbers of couples. The MN data have been rigorously scrutinised to exclude possible distortions due to erroneous typing (e.g. Reed and Milkovich 1968), and there have been a number of studies comparing observed gene frequency diversity locally in comparison with population structure (e.g. Friedlander 1971 a, b; Langaney 1972; Harpending and Jenkins 1973; Morton *et al.* 1971). There have been concerted attacks on the associations of the blood groups, particularly ABO, with specific diseases (e.g. Vogel 1973). A particularly interesting trend is manifested in the various attempts since 1967 to estimate how much selection may be present yet not detected by the data and method used (Neel and Schull 1968; Sing *et al.* 1971) from which it appears that the largest studies to date have been too small to detect some selective effects as large as 10%.

INITIALLY Professor E. B. Ford (1945) predicted that the allelomorphs controlling the ABO blood groups, the presence or absence of the ABH antigens in the saliva and the ability or inability to taste weak solutions of phenyl-thio-urea are all subject to natural selection. Since these genes were known not to affect appreciably either gross morphology or mating preferences (the choice of a particular husband or wife), such a prediction as the one made by

Ford could only imply that the various genotypes differ in fertility, in susceptibility to disease or in both. In fact, with regard to the blood groups and secretor character (presence or absence of ABH antigens in the saliva) Ford was fairly specific as to what associations should be looked for. He said, "A valuable line of enquiry which does not yet seem to have been pursued in any detail would be to study the blood-group distributions in patients suffering from a wide variety of diseases. It is possible that in some conditions, infectious or otherwise, they would depart from their normal frequencies, indicating that persons of a particular blood group are unduly susceptible to the disease in question. It is highly unlikely that the effect would be sufficiently large to suggest methods of treatment, though this is at least conceivable." When discussing the secretor character he again indicates the line of enquiry which, in his opinion, might prove fruitful. He writes, "It is not unlikely that it is ultimately concerned in the complex balance which is probably attained in the distribution of the O, A, B types, and it may play a part, which deserves investigation, in the relationship between mothers and embryos who belong to different blood groups of that series". Ford's approach to the problem of the association between hereditary factors and disease was biological rather than purely medical and it seems appropriate, therefore, to examine the validity of his predictions, at a symposium held by this Society, some fourteen years after Ford's paper was sent to press.

Aird, Bentall and Fraser Roberts (1953) found an association between blood group A and cancer of the stomach by using precisely the approach suggested by Ford. The next year, Aird, Bentall, Mehigan and Fraser Roberts (1954) reported that there was an association between group O and peptic ulceration. This was confirmed by Clarke, Cowan, Edwards, Howel-Evans, McConnell, Woodrow and Sheppard (1955) who also showed that there was a significantly greater association between group O and duodenal ulceration than there was with gastric ulceration. Since that time the difference between duodenal and gastric ulceration has been put beyond doubt, and it has also been found that there is an even greater association between stomal ulcers and group O than that found with the other two conditions. People with pernicious anaemia have been shown to have a high frequency of blood group A, and several other possible correlations have been reported. The whole subject has been reviewed by Fraser Roberts (1957).

The fact that the ABO blood groups were apparently connected in some way with one disease of the gastro-intestinal tract (since then the number has increased to four), led me to suggest (1953) that the secretor status of people with cancer of the stomach should be investigated. When in the next year the report on peptic ulceration was published we decided to investigate the blood group and secretor status of people with these diseases. Insufficient data concerning people with cancer of the stomach have been collected up to the present time for any worthwhile conclusions to be drawn, but we have shown a very strong correlation between the secretor status of a person and susceptibility to duodenal ulceration (Clarke, Edwards, Haddock, Howel-Evans, McConnell and Sheppard, 1956). This association has been confirmed independently by Wallace, Brown, Cook and Melrose (1958). Even more recently it has been possible to demonstrate that, as with blood group O, there is less association between ABH secretion and gastric ulceration than there is with duodenal ulceration. In fact, on the data available at the moment we have been unable to prove any effect of secretion on gastric ulceration (Clarke, Price Evans, McConnell and Sheppard, 1959). The data also give no indication of any interaction between the ABO blood groups and the secretor character on duodenal ulceration. That is to say, the increased liability to ulceration of a non-secretor group O person over one who is a non-secretor of groups A, B or AB, is the same as that of a secretor group O person over one who is a secretor of groups A, B or AB. This result shows that the increased susceptibility of group O people to duodenal ulcers is not the result of the H substance secreted into the gastro-intestinal tract being less protective or more ulcerogenic than the A or B substances since an equal effect of group O is found in non-secretors, where A, B and H substances are absent from the watery secretions of the body. By suitable biochemical investigations we have also been able to supply evidence which suggests that the effect of secretion and non-secretion is not due solely to the presence of more or less blood group substance in the mucus, since secretors and non-secretors seem to have about the same amount of blood group material present in their saliva—the only difference between them being that in non-secretors the blood group substance is Lewis, not ABH specific. The ABO and secretor characters do not appear to interact with one another with respect to ulceration. However, both genes control antigenic effects which are very closely associated, since one locus determines the type of antigen present in the body, whereas the other determines whether the particular antigen formed appears in the watery secretions of the body or not. Now both these loci also affect susceptibility to ulceration and it therefore seems unlikely that they act by way of pleiotropic effects which are quite unrelated with one another. It seems much more reasonable to think that they are each producing their effects on ulceration in a similar manner, probably by way of some complex antigenic effect.

Recent clinical findings (see Rosenfield, 1955) show that the ABO system can be responsible for haemolytic disease of the new-born, particularly when the mother is group O. One would expect that any incompatible foetal red-cells entering the maternal circulation would be destroyed by her antibodies before there was time for her to be stimulated to manufacture immune Anti–A or Anti–B against the foetal antigen. That this view is in part correct is indicated by the work of Smith (1945) and others who have shown that most infants suffering from ABO haemolytic disease are secretors of ABH substance, thus suggesting that immunization is usually by way of water soluble antigens and not by way of the red cells.

The interactions between the human polymorphic characters are even more complex than I have so far indicated. It has been suspected for some time (see Levine, 1958) that in certain circumstances the ABO blood groups afford protection against erythroblastosis foetalis due to Rh incompatibility between mother and foetus. It has been suggested that, because of a rapid destruction of foetal red-cells incompatible with the mother on the ABO system, an Rh–positive ABO incompatible foetus will not sensitize its Rh–negative mother against Rh. However, once sensitization has occurred, subsequent pregnancies can give rise to erythroblastosis since even ABO incompatible cells are not destroyed sufficiently quickly to prevent a sensitized mother from making Rh antibodies. Clarke, Finn, McConnell and Sheppard (1958) have investigated this hypothesis by blood-grouping the parents and children of families in which Rh haemolytic disease of the newborn has occurred. Of 22 suitable families (those in which the parents' genotypes are such that an ABO incompatible foetus could be produced) the sensitizing foetus was known in 14 cases, and in each instance its ABO blood group was compatible with the mother on the ABO system, a result which would not be expected by chance as often as once in five hundred trials. Moreover, in the remaining eight families, in which the sensitizing foetus was not known for certain, at least one ABO compatible child was born before Rh erythroblastosis made its appearance. Thus our data strongly support the hypothesis and are not inconsistent with an ABO incompatible foetus giving complete protection against Rh sensitization of the mother. After sensitization has occurred no as yet detectable protection is afforded, and a child of any ABO phenotype can develop Rh haemolytic disease of the newborn. Besides the data given above, there are recorded two families which might have been considered as exceptions since the disease appeared before an ABO compatible child had been born; however, they had to be excluded from the survey since in both instances the mother had previously had a blood transfusion (before 1944), which was probably the sensitizing agent. The conclusion that the ABO blood groups can protect against Rh sensitization (and almost certainly Kell—see Race and Sanger, 1958) is further strengthened by the work of Stern, Davidsohn and Masaitis (1956) who attempted to induce Rh antibody formation in male Rh–negative volunteers by the injection of Rh–positive blood. Their success was greater when the blood was ABO compatible with the recipient than when it was not.

We can see that Ford (1945) was correct in believing that associations between the ABO blood groups and disease would be found, that there would be some interaction between the ABH secretor and the ABO blood group phenotypes, and that in this respect it was important to investigate the secretor status of mothers and embryos who differed in their ABO blood group.

His third suggestion, namely, that the ability or inability to taste phenyl–thio–urea (and some related chemical substances) would be subject to natural selection has also born fruit. Harris, Kalmus and Trotter (1949), knowing that the substances which distinguish tasters from non-tasters also affect thyroid function, decided to investigate the frequency of non-tasters among people with goitre. They found a doubtfully significant excess of non-tasters among patients with nodular goitre. We have carried out a similar investigation at Liverpool (Kitchin, Howel-Evans, Clarke, McConnell and Sheppard, 1959) and found not only a significant excess of non-tasters among patients, both with toxic and non-toxic multiple adenomata, particularly in the males, but also an equally striking deficiency of them among people with toxic diffuse goitre. Thus again Ford's prediction has been verified.

In view of the accuracy of these predictions it seems worth while discussing briefly the considerations which led Ford to make them. He said: "It is reasonable to conclude, from what we know of polymorphism, that individuals belonging to the different blood groups are not equally viable". It was Fisher (1930) who showed that the effects of a single gene can seldom be neutral in survival value for any appreciable length of time, particularly if they be large. Moreover, even if such a gene existed its spread in a large population would be extremely slow. Consequently, "If a genetically controlled variety occupies even a few per cent of a population, we can be fairly sure that it possesses some advantages" (Ford, 1945). Furthermore, if two or more forms can be shown to have persisted for a very long time at such a frequency, the polymorphism is likely to be a stable one and, therefore, the advantages of each allelomorph must be counterbalanced by some equal disadvantage. A common way for this to be brought about is by the heterozygote being at an advantage to both of the homozygotes (Fisher, 1930), as has been explained to us by Dr. Owen (p. 27). Now in the case of the ABO blood groups and the ability or inability to taste phenyl–thio–urea, we have evidence that the polymorphism is older than Man himself. Some of the great apes are polymorphic for ABO blood group antigens (chimpanzees have A and O; orang-utans and gibbons have A, B and AB—see Mourant, 1954), and both chimpanzees and orang-utans are polymorphic for the ability or inability to taste phenyl–thio–urea as was shown by Fisher, Ford and Huxley (1939). Consequently, we are fairly safe in predicting that the polymorphism is stable and that we will find advantages and disadvantages attributable to the characters controlled by the allelomorphs concerned. However, the exact conditions necessary to establish a stable polymorphism when there are more than two allelomorphs, as with the ABO blood groups, may be complex as Dr. Owen has pointed out. Moreover, since we know that the interactions between the ABO, Rh and secretor loci are far from simple themselves, it may be difficult to formulate a mathematical hypothesis to account for the persistence of the various phenotypes in human populations. Nevertheless, from what has already been said, I think we can infer three things:

1. The human polymorphisms characterized by the blood groups, serum haptoglobins, haemoglobins and many others, will be found to affect fertility or susceptibility to disease.

2. Their effects will be of a complex nature. For example, the heterozygote for the gene controlling sickle-cell haemoglobin is known to be protected to some extent against malaria and the homozygote to suffer from sickle-cell anaemia (Allison, 1956/7), thus allowing the polymorphism to develop and be maintained in malarial areas. However, it is more than likely (by analogy with other polymorphisms) that this gene and its allelomorphs are controlling several characters besides those already known and that these also will have important effects on disease and fertility.

3. The selection pressures acting on the effects of the genes controlling the ABO, Rh, Secretor and taste polymorphisms, discussed today, do not appear to be of sufficient magnitude to account for, nor of such a nature as to lead to stable polymorphism.

We are, therefore, led to the conclusion that we must look for other and selectively more powerful effects which must be so balanced that they will result in a stable gene-frequency equilibrium. I have discussed this problem elsewhere (Sheppard, 1959) but nevertheless, will mention a few points which seem to me important.

The most likely explanation for the persistence of two or more allelomorphs in a population, as we see in the human polymorphisms mentioned, would seem to be that the heterozygotes are in some way at an advantage, as was mentioned earlier by Professor Penrose. Now it has already been shown that this supposition is true in the case of at least some chicken blood groups

(Schultz and Briles, 1953; Briles, Allen and Millen, 1957; Gilmour, 1954, 1958) and Dr. C. Cohen tells me that all the quite small populations of domestic rabbits investigated by him have always been found to be polymorphic for at least one blood group system. This again suggests that heterozygotes are at an advantage since, in at least some small inbred populations, one would expect homozygosity to be attained by inbreeding alone if selection were not operating.

In man, it has been noticed that there is often an excess of MN heterozygotes among the children of parents themselves heterozygous for this blood group system. This is accounted for in some data by faulty antisera being used since in such families all three genotypes would be expected and, therefore, mistakes would not necessarily be noticed. However, many investigators also give data for population surveys which act as a control since, if the antisera or technique were at fault, there should be a similar excess of heterozygotes in the population data, as well as among the parents of the families tested. In fact, even if we exclude data in which such an excess occurs in the population control, we still get more heterozygous children from heterozygous parents than would be expected. Thus there is some evidence that there is selection in at least one human blood group system of such a kind that it will give rise to a stable polymorphism. However, in view of the fact that some large surveys show no such excess of heterozygous offspring (see Race and Sanger, 1958), the presence of the selection cannot yet be considered as established.

There is a mechanism known which could account for an increase in the frequency of Rh–negative people despite the disadvantage resulting from erythroblastosis foetalis. Fisher pointed out that if the parents of families in which Rh haemolytic disease occurred tended to produce children to replace those who died, they might overcompensate and, in the end, have more living children on average than unaffected families. Such a situation could lead to an increase in the frequency of Rh–negatives in a population, and Glass (1950) demonstrated the presence of such compensation in the American white population but not in negroes where contraception is less often practised. However, under constant conditions, this compensation will not lead to a polymorphism which is stable (Li, 1953). In fact, below a certain critical gene-frequency, the value of which depends on the magnitude of the compensation effect, Rh–negatives will still tend to decrease in frequency, but above it they will tend to completely replace Rh–positives in the population. Consequently, even in the case of the Rh and Kell systems, both of which can be responsible for erythroblastosis, we need some other selective effect to account for the persistence of polymorphism.

The ABO polymorphism is equally difficult to explain since it also can cause haemolytic disease of the newborn. It is true that in Western Europe the interaction between this system and Rhesus will help to preserve the stability because it will put group B at an advantage while it is rare, and at a disadvantage compared with group A should it become common. This follows from the fact that the commoner the gene the smaller will be the proportion of B foetuses which are ABO incompatible with their mother and protected from sensitizing her against Rh. However, since only about one child in one hundred and fifty suffers from Rh erythroblastosis the selection pressure involved will be too small to have much effect in the face of counter-selection due to ABO haemolytic disease. There are many reports of very high elimination rates of foetuses which are incompatible with the mother on the ABO system but by no means all workers have found data which support this view (see Sheppard, 1959) and, until more adequately controlled data are available, judgement on the selective importance of ABO incompatibility must be suspended. There is also some evidence that the secretor character, or some variant of it, plays a part in this elimination (McNeill, Trentleman, Fullmer, Kreutzer and Orlob, 1957).

In order for rapid progress to be made in the field I have been discussing today, it seems essential to try to get a reasonable estimate of the intensity of the selection acting on the human polymorphisms. There seem to be several lines of approach which could be adopted and I propose to mention four which I think are worthy of special attention.

1. The effects of the ABO system and secretor character (including the so-called aberrant secretor) on fertility, abortion rate and sex-ratio should be reinvestigated in detail, preferably in a rather homogeneous and static community when the difficulty of obtaining adequate controls is more easily overcome. Such an investiga-

tion should embody as complete an assessment of the genotype of each individual as possible in order to eliminate errors due to illegitimacy (for the importance of this see Edwards, 1957).

2. The apparent excess of MN heterozygous offspring of matings between heterozygotes should be reinvestigated using rigid controls which would exclude the possibility of erroneous results due to incorrect typing.

3. A possible advantage of blood group heterozygotes in mammals similar to that found in chickens should be investigated in several species. In laboratory animals this could be done by suitable breeding techniques. In man it might be accomplished by a forward survey along the lines of the work of Reed and Kelly (1958). By selecting a large number of newly-married couples and recording their subsequent reproductive performance together with a complete assessment of the blood groups, serum proteins, secretor status and taste response of all concerned, it should be possible in the future to answer many of the problems connected with the intensity of selection on human polymorphic characters.

4. A population survey should be made in order to ascertain, if possible, the relative importance of selection and genetic drift in determining the difference between human populations in respect of polymorphic characters. Mourant (1954) has pointed out that although the frequencies of allelomorphs at the ABO locus vary markedly from place to place, those for some other blood group systems do not show so great a tendency to vary over short distances (e.g., see Ceppellini, 1955). If all the differences between populations were due to genetic drift one would expect about the same degree of variation in all blood group systems. Consequently, the data suggest that quite powerful selective forces are acting on at least some systems. By making a complete survey of the breeding structure of a population such as that of Cavalli-Sforza (1959) or of Roberts (1956/7), one could determine the amount of diversity in gene-frequency from neighbourhood to neighbourhood which could be expected if all of it were due to genetic drift. Then by adequate sampling of the population for blood group frequencies and other polymorphic traits one could determine whether the observed variation could reasonably be ascribed to chance alone. Any variation too great or too small to be accounted for by drift

would indicate the action of natural selection, and might give some estimate of the minimum selection pressures involved. The main disadvantage of such an approach would be that variation consistent with drift could, nevertheless, result from selection varying in direction and intensity from place to place. This could be partially offset if several surveys were made, since it would be unlikely that all would be consistent with drift if there were strong selective forces operating.

Many independent lines of enquiry have shown that human polymorphisms are subject to natural selection. The time has now come for a systematic attack on such problems as (i) the methods by which the polymorphisms are maintained, (ii) the strength of the selection involved and its variation in magnitude from place to place, (iii) the advantages and disadvantages associated with the various polymorphic systems, and (iv) the methods by which the genes exert their important effects. The last problem, if solved, might suggest methods of treatment or prevention of the diseases concerned as Ford (1945) pointed out.

Acknowledgements

I am very grateful to Dr. C. A. Clarke, Dr. E. B. Ford, F.R.S., and Dr. G. A. Harrison for reading the manuscript and for their helpful comments. I also wish to express my thanks to the Nuffield Foundation for their generous support of the work on polymorphism.

References

AIRD, I., BENTALL, H. H. and ROBERTS, J. A. F. (1953) A relationship between cancer of the stomach and the ABO blood groups. *Brit. Med. J.* **1**, 799.

AIRD, I., BENTALL, H. H., MEHIGAN, J. A. and ROBERTS, J. A. F. (1954) The blood groups in relation to peptic ulceration and carcinoma of the colon, rectum, breast and bronchus. *Brit. Med. J.* **2**, 315.

ALLISON, A. C. (1956/57) The sickle-cell and haemoglobin *C* genes in some African populations. *Ann. Hum. Genet.* **21**, 67.

BRILES, W. E., ALLEN, C. P. and MILLEN, T. W. (1957) The B blood group system of chickens. 1. Heterozygosity in closed populations. *Genetics* **42**, 631.

CAVALLI-SFORZA, L. L. (1959) Some data on the genetic structure of human populations. *Proc. 10th Int. Congr. Genetics* **1**, 389.

CEPPELLINI, R. (1955) In discussion of "Aspects of polymorphism in man". *Symp. Quant. Biol.* **20**, 252.

Clarke, C. A., Cowan, W. K., Edwards, J. W., Howel-Evans, A. W., McConnell, R. B., Woodrow, J. C. and Sheppard, P. M. (1955) The relationship of the ABO blood groups to duodenal and gastric ulceration. *Brit. Med. J.* 2, 643.

Clarke, C. A., Edwards, J. W., Haddock, D. R. W., Howel-Evans, A. W., McConnell, R. B. and Sheppard, P. M. (1956) ABO blood groups and secretor character in duodenal ulcer. *Brit. Med. J.* (ii), 725.

Clarke, C. A., Finn, R., McConnell, R. B. and Sheppard, P. M. (1958) Protection afforded by ABO incompatibility against erythroblastosis due to Rh and anti–D. *Int. Arch. Allergy, N.Y.* 13, 380.

Clarke, C. A., Price-Evans, D. A., McConnell, R. B. and Sheppard, P. M. (1959) The secretion of blood group antigens and peptic ulcer. *Brit. Med. J.* 1, 603.

Edwards, J. H. (1957) A critical examination of the reputed primary influence of ABO phenotype on fertility and sex ratio. *Brit. Med. J. Prev. Soc. Med.* 11, 79.

Fisher, R. A. (1930) *The Genetical Theory of Natural Selection.* Clarendon Press, Oxford.

Fisher, R. A., Ford, E. B. and Huxley, J. S. (1939) Taste-testing the Anthropoid Apes. *Nature, Lond.* 144, 750.

Ford, E. B. (1945) Polymorphism. *Biol. Rev.* 20, 73.

Gilmour, D. G. (1954) Selective advantage of heterozygosis for blood group genes among inbred chickens. *Heredity* 8, 291.

Gilmour, D. G. (1958) Maintenance of segregation of blood group genes during inbreeding in chickens. *Heredity* 12, 141.

Glass, B. (1950) The action of selection on the principal Rh alleles. *Am. J. Hum. Genet.* 2, 269.

Harris, H., Kalmus, H. and Trotter, W. R. (1949) Taste sensitivity to phenylthiourea in goitre and diabetes. *Lancet* 2, 1038.

Kitchin, F. D., Howel-Evans, A. W., Clarke, C. A. McConnell, R. B. and Sheppard, P. M. (1959) The P.T.C. taste response and thyroid disease. *Brit. Med. J.* 1, 1069.

Levine, P. (1958) The influence of the ABO system on Rh hemolytic disease. *Hum. Biol.* 30, 14.

Li, C. C. (1953) Is Rh facing a crossroad? A critique of the compensation effect. *Am. Naturalist* 97, 257.

McNeill, C., Trentleman, E. F., Fullmer, C. D., Kreutzer, V. O. and Orlob, R. B. (1957) The significance of blood group conflict and aberrant salivary secretion in spontaneous abortion. *Am. J. Clin. Path.* 28, 469.

Mourant, A. E. (1954) *The Distribution of the Human Blood Groups.* Blackwell, Oxford.

Race, R. R. and Sanger, R. (1958) *Blood Groups in Man,* 3rd ed. Blackwell, Oxford.

Reed, T. E. and Kelly, E. L. (1958) The completed reproductive performances of 161 couples selected before marriage and classified by ABO blood group. *Ann. Hum. Genet. Lond.* 22, 165.

Roberts, D. F. (1956/7) Some genetic implications of Nilotic demography. *Act. Genet. Stat. Med.* 6, 446.

Roberts, J. A. F. (1957) Blood groups and susceptibility to disease. *Brit. J. Prev. Soc. Med.* 11, 107.

Rosenfield, R. E. (1955) AB hemolytic disease of the newborn. Analysis of 1480 cord blood specimens, with special reference to the direct antiglobulin test and the group O mother. *Blood* 10, 17.

Schultz, F. T. and Briles, W. E. (1953) The adaptive value of blood group genes in chickens. *Genetics* 38, 34.

Sheppard, P. M. (1953) Cancer of the stomach and the ABO blood groups. *Brit. Med. J.* 1, 1220.

Sheppard, P. M. (1959) Blood groups and natural selection. *Brit. Med. Bull.* **15,** 134.

Smith, G. H. (1945) Iso-agglutinin titres in heterospecific pregnancy. *J. Path. Bact.* **57,** 113.

Stern, K., Davidsohn, I. and Masaitis, L. (1956) Experimental studies on Rh immunization. *Amer. J. Clin. Path.* **26,** 833.

Wallace, J., Brown, D. A. P., Cook, I. A. and Melrose, A. E. (1958) The secretor status in duodenal ulcer. *Scot. Med. J.* **3,** 105.

ON SELECTION OF GENE SYSTEMS IN NATURAL POPULATION

THEODOSIUS DOBZHANSKY

Here there has emerged an interesting dichotomy. Undoubtedly, the importance of co-adapted gene complexes has been well established in many organisms, particularly drosophila (e.g. Dobzhansky 1966). The theoretical evolutionary advantage that this confers has been well established mathematically (e.g. by the simulation work of Holland 1973). Yet in man apart from a few situations of apparently close linkage as in the rhesus blood groups, there is very little evidence that human populations are differentiated in such complexes.

PROBABLY all of us have learnt genetics, and many have also taught it, in terms of genes with clear-cut, discrete, and easily recognizable effects. Such, indeed, were the genes which Mendel, Bateson, and Morgan chose to study; such are most of the genes discussed in treatises on medical genetics. The concentration of the attention of investigators on "major" genes with discrete phenotypic effects was wise strategy when these genes were used to study linkages, chromosomal aberrations, and the physiology of gene action; it was also wise strategy when the problem was to convince people that genetics had some predictive value when applied to man. Pioneers of genetics were not unaware of the existence of more complex forms of transmission of heredity but they elected to leave these temporarily in abeyance. For example, Morgan (1931) had the following to say concerning the inheritance of human stature: ". . . it may seem that the best we could do is to ignore these complications and treat stature as a statistical problem. Such treatment would not furnish readily an analysis of the factors concerned. On the other hand, it may be only a matter of time to find out how many genetic factors are present, and then to isolate those that determine growth." The complications could no longer be ignored when it became clear that the characteristics of organisms most important in the evolutionary process, and also in the agricultural breeding work, prove usually to have an intricate genetic basis. Elaboration of methods of treating the inheritance of such characteristics "as a statistical problem" should be credited before all to Mather and his colleagues in Birmingham, to the Edinburgh school of statistical genetics, to Lerner in California, to the Penrose school in London, and to other investigators elsewhere.

The focusing of the attentions of geneticists on "major" genes had yet another, and less desirable, effect. Genetics has demonstrated that heredity is particulate; in Morgan's words, "There are elements in the germ-plasm that are sorted out independently of one another". Although Morgan saw clearly that ". . . the whole germ-plasm, the sum of all the genes, acts in the formation of every detail of the body . . .", the semantic habits of geneticists conspired to conceal this fact. We still speak of the gene "for" the blood group A, the haemophilia gene, the achondroplasia gene in man, the gene "for" resistance to DDT in the house fly, and "for" resistance to the newest antibiotic in bacteria. Recent work in genetics has shown that the atomism of heredity is actually not as thorough-going as it was believed earlier. Anyway, the process of individual development, of the realization of heredity in the ontogeny, is unitary.

The form of discourse accepted in genetics is liable to generate pseudo-problems. What, for example, was the gene for DDT resistance doing in the house flies during the aeons of time before DDT was discovered? If evolution is directed by natural selection, why is it so often difficult to discover the adaptive significance of so many "characters" and "traits" in man and in other organisms? The difficulties might become less formidable if it were recognized that ontogeny does not consist of gradual accretion of autonomous characters, each engendered by a separate gene. The development is brought about by the genotype, the constellation of genes, acting in concert, and interacting with the environment. There is no gene "for" DDT resistance in flies, although the difference between the resistant and the non-resistant flies may be conditioned by one gene, or by a small group of genes. A trait or a character is not a part of the body prefabricated by a single gene; it is rather an aspect of the development pattern of the whole organism. Some development patterns confer high fitness in some environments, and other patterns in other environments. Still other development patterns are adaptively incompetent in all existing environments; they are referred to as incurable hereditary diseases.

Morgan's dictum that all the genes which an organism possesses participate in the formation of every detail of the body raises the problem of how the genotypes of individuals and of populations are organized to yield adaptive traits and adaptive development patterns. With some genes the problem is solved easily—they seem to affect a single organ or a single trait, are fully penetrant over the entire range of the environments in which the species lives, and their presence or absence is readily recognizable in the phenotype. Such are the "good" genes of classical genetics, and also the genes which produce hereditary diseases and malformations in man which are used as illustrations in textbooks of general and of human genetics. For such genes it is possible, at least in principle, to estimate the selection and mutation pressures which maintain them and their alleles in the population at observed frequencies, and the numbers of "genetic deaths" which they produce within a given time span.

More complex situations arise when the genetic basis of an adaptive trait is an array of genes with individually small effects (polygenes). The complexity is greatest when the action of a given gene depends upon what other genes are included in the same genotype (epistatic gene interactions). The density of such a gene in populations under selection will evidently depend on the fitness of the array of genotypes in which this gene is combined with genes at other loci present in the gene pool. For what natural selection will be doing will be to further or to impede the spread of gene systems, rather than of single genes, adaptive in certain environments. It becomes unrealistic to speak of a given gene being favourable or unfavourable to the species as a whole, since this allele may depress the fitness of some, and enhance that of other, genotypes.

It is tempting to speculate that the effects of genes on fitness tend to be autonomous more frequently in lower organisms, while in higher and more complex organisms epistatic interactions are more prevalent. Natural selection will, then, manipulate single genes most frequently in lower, and gene systems in higher organisms. How great a variety of adaptive different genotypes may arise owing to recombination of genes present in the gene pool of a single population of a species remains quite insufficiently known. I would like to summarize briefly the results of experiments bearing on this issue, carried out in three species of Drosophila (*D. pseudoobscura, D. persimilis,* and *D. prosaltans*) by Spiess, Levene, Spassky, and the present writer.

When the genetic composition of a natural population of Drosophila is analysed, it is usually found that considerable proportions of their chromosomes (other than the heterochromosomes) are deleterious to their carriers when present in double dose (in homozygous condition). Anywhere from 10% to as many as 60% of the "wild" chromosomes are lethal or semi-lethal when homozygous; among the remainder, a majority, up to 95%, of the chromosomes are subvital when homozygous, i.e., produce relatively mild constitutional weakness. Some 10% to 20% of the non-lethal and non-semi-lethal chromosomes act as recessives which make females completely sterile, and a fairly similar proportion sterilizes the male homozygotes. Recessive effects inducing pronounced morphological changes in the homozygotes are relatively rare, less than 5% of the chromosomes.

Now, in each of the three species named above we chose 20 chromosomes from natural populations; these chromosomes yielded normally viable, or only slightly subvital, homozygotes. By means of a suitable system of crosses, all possible combinations of the chromosomes were obtained—20 chromosomes yield 190 combinations. Females carrying each pair of the chromosomes were outcrossed to mates with appropriate gene "markers"; 10 sons were taken from each progeny, a total of 1,900 males. Each of these males carried a chromosome which was likely to be a product of recombination by crossing-over between the two "wild" chromosomes in the mother. Other crosses were then made to test the viability effects of each of the 1,900 recombination chromosomes in homozygous condition. Although the original chromosomes yielded normally viable homozygotes, among the recombination chromosomes there was a great variety of viability conditions, ranging all the way from complete lethality ("synthetic lethals"), semi-lethality, subvitality, to normality, and even to supervitality.

In *D. pseudoobscura,* 77 out of 1,900 recombination chromosomes, or about 4%, were lethal. The corresponding figures for *D. persimilis* and *D. prosaltans* were 2·4% and 5·7%.

The appearance of synthetic lethals means that recombination of genes contained in two "healthy" chromosomes may produce chromosomes which cause fatal hereditary diseases when homozygous. It should, however, be noted that synthetic lethals are merely extreme situations in which gene recombination yields gene systems with unexpected properties. It is even more important that the total amount of the genetic variance released by recombination in our experiments proved to be very large. We believe that the fairest comparison of the variance released by recombination is with the total amount of the genetic variance present among the chromosomes in natural populations of the same species. The relevant figures for the variance present in nature ("wild chromosomes"), and for the variance released by recombination are as follows:

Species	Wild Chromosomes	Recombination Chromosomes	Percentage Ratio
D. pseudoobscura	140	60	43
D. persimilis	110	26	24
D. prosaltans	200	50	25

It can be seen that the average amount of the genetic variance released by recombination, in a single generation, of the gene contents of wild chromosomes selected for uniformity and "normality" comes up to between 24% and 43% of the total genetic variance available in natural populations of the species studied. To express the same result differently: the maintenance of the genetic variability in populations of species with genetic structures like those in the Drosophila species studied depends on recombination to a greater extent than it does on recurrent mutation. If the process of mutation were temporarily suppressed, the amount of the genetic variance, and hence the evolutionary plasticity, might remain unaffected for a considerable time.

An assumption which has gained widespread credence among evolutionists is that high mutation frequencies in a species should be conducive to rapid evolution, and, conversely, that rarity of mutation would tend to evolutionary conservatism. This assumption need not necessarily be valid, at least not in all forms of life.

The evolutionary patterns may well be different in haploid and in diploid organisms, in those reproducing mainly or exclusively asexually and in the obligatory sexual and cross-fertilizing ones, in those in which epistatic interactions of genes occur but rarely and in those in which they are prevalent. One is tempted to speculate that the evolutionary plasticity may show a pronounced positive correlation with the mutability in bacteria and in viruses; among higher organisms the evolutionary rates should depend more upon the intensity and the kind of natural selection, rather than upon mutation rates. One must, however, guard against underestimation of the complexity of the situation.

We may be forced to stress more than the classical geneticists did the differences between the evolutionary roles of major mutants and of the polygenic variability. Thus, it seems probable that the maintenance in human populations of a certain incidence of hereditary diseases and anomalies, such as haemophilia or chondro-dystrophy, may be due to the recurrent mutation pitted against selection. On the other hand, the all-pervading, if less striking, differences between "normal" men may be due to the polygenic variability maintained by selection rather than by mutation. Distinguishing these kinds of variability does not necessarily imply a dualism in the genetic materials, or for that matter in the kinds of mutational changes that occur.

Little is known about the forms which natural selection may take when it operates with gene systems, rather than with genes the manifestation of which is effectively autonomous from the rest of the genotype. Studies on the genetics of natural populations of Drosophila have disclosed a number of interesting situations; some of them may be briefly described here, even though it is unlikely that they bear very close resemblance to what may be found in the evolution of man. Natural populations of many species of Droso-phila are mixtures of two or several chromosomal types differing in inversions of blocks of genes. Since flies with different chromo-somes interbreed freely, individuals found in nature may have the two chromosomes of a pair of the same type (structural homo-zygotes) or of different types (structural heterozygotes). Each type of chromosome occurs in populations of a species in a certain geographic region; some chromosomal forms are widespread and others more limited in distribution. Experimental populations of

Drosophila can be created in the laboratory and they can be made to contain any desired proportions of two or more chromosomal types. The flies with different chromosomes are, however, unequally fit to live in the experimental populations. Natural selection in the experimental populations acts accordingly; the incidence of some chromosomal types rises, and of others declines, from generation to generation.

The outcome of the experiments is simplest if the chromosomal types in an experimental population are of geographically uniform origin, i.e., derived from ancestors collected in the same geographic locality. In *Drosophila pseudoobscura,* which is the species used most extensively in such experiments, the highest fitness is exhibited usually by heterozygotes, having the two chromosomes of a pair of different type but of the same origin geographically. The homozygotes have a lower fitness. Because the heterozygotes are heterotic, natural selection establishes genetic equilibria in the experimental populations. The two or more competing chromosomal types are preserved in the populations, each with an equilibrium frequency which usually remains constant for as long as the experiment lasts.

A most remarkable difference in the outcome of experiments on laboratory populations of Drosophila is observed depending upon whether the chromosomal types which are made to compete in these populations are derived from progenitors taken in the same or in different geographic localities. Thus, in an experiment of Dobzhansky and Pavlovsky four replicate experimental populations were started with 20% of the chromosomes of the type denoted as ST and 80% of the type CH, all chromosomes derived from flies collected in a certain locality in California. The frequencies of ST chromosomes rapidly rose, and those of CH chromosomes fell, in every population. A year later (about 15 fly generations), all populations contained 80–84% of ST, and 16–20% of CH chromosomes. Moreover, the rates of increase of the frequencies of ST chromosomes in the four replicate populations were alike within the expected limits of sampling errors.

Something else has, however, happened in another set of four experimental populations, which differed from the first set only in that the ST chromosomes were derived from flies collected in California, while the CH chromosomes came from a locality in

Mexico. As in the first set, the frequencies of ST chromosomes rose and those of CH declined. However, the progress of selection was different in every population. A year after the start, one of the populations had only 80% of ST chromosomes, while the three other populations had 97–99%. Furthermore, in some populations the rate of increase of the frequency of ST was more rapid than in others. In short, the four replicate populations with a mixture of chromosomes of Californian and of Mexican origin failed to give uniform results; every population behaved differently from the others.

But why should the experiments with populations of uniform geographic origin give more reproducible results than those of mixed origins? In what do the situations in these two kinds of populations differ? Consider the populations of geographically uniform origin first. The chromosomes in these experimental populations existed together in the same natural population; their gene contents were mutually adjusted, or coadapted, by natural selection to produce heterotic heterozygotes on the genetic background of the population in question; the action of the natural selection observed in our experimental population is, then, merely causing the chromosomal types to find the predetermined frequency levels which confer the highest fitness on these populations.

Considerably more complex processes are enacted in the experimental populations which contain mixtures of chromosomes of diverse geographic origins. The flies in these experimental populations are hybrid progenies from crossing different geographic races—a race from California and a race from Mexico. These hybrids are entirely artificial products, in the sense that such hybrids presumably never occurred in nature (the probability of a fly migrating from California to Mexico, or vice versa, is negligible). In any case, a large variety of genotypes must arise in the populations following the interracial hybridization. We do not know in how many genes the races differ, and as a consequence cannot estimate, even approximately, the number of new gene combinations which may, potentially, be formed in the hybrid populations in the process of gene assortment. One thing which is certain is that this number must be vastly greater than the numbers of individuals from which the experimental populations consist (1,000–4,000). Now, natural selection will not remain inactive until the fittest of all the potentially possible genotypes is produced; it will augment the frequency of any relatively fit array of genotypes which happens to arise first in a given population. Different genotypes will probably arise in different populations. The divergent results in replicate experiments, in apparently identical populations kept in identical environments, are a consequence. Such, at least, is the working hypothesis to be tested.

The following test was devised (Dobzhansky and Pavlovsky, 1957). The hypothesis postulates that the varying results of the selection process in the experimental populations arise from a disproportionality between the population size and the capacity of the sexual process to generate new gene combinations. If so, the outcomes of selection should be more variable in smaller populations, and less so in larger ones. The operation of selection would become presumably wholly determinate in ideal infinite populations. We have accordingly made 20 experimental populations, all descended from F_2 generation hybrids between flies from California carrying a chromosome of the type denoted AR, and flies from Texas with a chromosome called PP. However, 10 of these populations were descended from groups of 20 flies taken at random from the F_2 hybrid progenies; the other 10 populations were each descended from 4,000 "founders" of the same origin.

Although some of the populations came from small and others from relatively large groups of founders, so great is the fecundity of the flies that all the experimental populations grew equally large one generation after the start. About 17 months (some 20 to 21 generations) later the populations descended from small numbers of founders contained from 16–47% PP chromosomes. Those descended from larger numbers of founders had from 20–35% PP chromosomes. The "small" populations gave, indeed, more variable results than the "large" ones. The variance of the observed frequencies in the latter is about 27, and in the former about 119, or 4·4 times larger. The difference is statistically significant.

The experiments now under way are giving further verification of the hypothesis. We have set up ten new experimental populations with AR chromosomes from California and PP chromosomes from Mexico. All the populations are descended from groups of 20 founders; the populations are permitted to expand, but at intervals of about four months each population is reduced to a sample of

20 founders, and permitted to expand again. Moreover, in five of these populations the original foundation stock came from mixing together a dozen different Californian and a dozen Texan strains. In the remaining five populations the founders were hybrids of only a single Californian and a single Texan strains. Natural selection has acted in all these populations, and a significant heterogeneity of the outcomes has appeared in both series. The heterogeneity is, however, strikingly greater among the populations which had genetically more heterogeneous foundation stocks.

While the evolutionary changes of the kind observed in the laboratory experiments are suggestive, they cannot be said to have proven that phenomena of the same kind are of importance also in evolution in nature. Fortunately, even before these experiments were completed, Mayr (1954) summarized a great deal of evidence from zoological systematics which solidly substantiates the validity of the inference. In species after species, and in different groups of animals, an interesting contrast is observed between the variability of continental and of island populations. The inhabitants of extensive but more or less continuously inhabited territories, such as continental masses, may show relatively little geographical differentiation. By contrast, populations isolated on islands, or by some distributional barriers, are often very appreciably different from each other and from the continental populations. Mayr has stressed the fact that environmental differences between parts of the continent may be much greater than those between the islands and the adjacent portion of the continent. The magnitudes of the racial differences are, thus, not proportional to the environmental differences in the territories which the races inhabit.

Mayr has pointed out that, despite environmental differences, genetic differentiation of continental populations may be impeded by migration and interchange of genes between the populations. However, populations isolated on islands, or by other barriers to migration, are more or less protected from the levelling effects of the interpopulational gene exchange. More important still, populations of islands or other distributional pockets are likely to be descended from single pairs, or from small groups, of migrants from the continent. These migrants brought with them not the entire gene pool of the parental population but only small segments thereof. The migrants are comparable to the small groups of the "founders" of our experimental Drosophila populations. Now, the gene pool of a Mendelian population is an internally balanced system. Natural selection will act in an island population to bring about a new balanced state, in place of the one disrupted by the sudden shrinkage of the gene pool in the founding population. This genetic reconstruction alone might be expected to cause the isolated population to become different from the continental one; any peculiarities of the island environments would, certainly, act as a further stimulus to differentiation.

It is admittedly unlikely that genetic situations just like those in our experimental Drosophila population will be found in natural populations of higher vertebrates or in man. Blocks of polygenes locked-up in chromosomal inversions seem to be important in natural populations mainly of dipterous insects and of certain plants. However, effectively similar "supergenes" may arise owing to a variety of other genetic processes. What really matters in the experiments described above is the demonstration that natural selection may work with organized gene systems, rather than with single discrete genes.

The elementary evolutionary events are changes in the incidence of genes in populations. The allele, or group of alleles, which was initially present in an ancestral population is gradually replaced by another allele, or a group of alleles. The replacement is due generally to natural selection, and consequently is likely to promote the adaptive value of the population in some environments. However, as pointed out by Wright (1955), "Each gene replacement inevitably has extensively ramifying pleiotropic consequences. In this situation genes that have favourable effects at all will also in general have many more or less unfavourable effects, with the net effect dependent on the array of other genes. Evolution depends on the fitting together of favourable complexes from genes that cannot be described as in themselves either favourable or unfavourable. The consequence of this situation is that there is not one goal of selection, but a vast number of distinct possible goals . . ."

References

DOBZHANSKY, TH. and PAVLOVSKY, O. (1957) An experimental study of interaction between genetic drift and natural selection. *Evolution* **11**, 311.

MAYR, E. (1954) Change in genetic environment and evolution. In *Evolution as a Process* (ed. by HUXLEY, HARDY and FORD), p. 157. Allen & Unwin, London.

MORGAN, T. H. (1932) *The Scientific Basis of Evolution,* Norton, New York.

WRIGHT, S. (1955) Classification of the factors of evolution. *Cold Spring Harbor Symp. Quant. Biol.* **20,** 16.

RATE OF CHANGE IN PRIMATE EVOLUTION

E. H. ASHTON

Of the major developments in this topic, first has been the great increase in the amount of fossil material available for representatives of the higher primates, and its accurate dating, primarily from Africa (Lake Rudolf, Omo, Olduvai, Baringo), but also from Asia (Siwaliks), so that the sequence of hominid emergence, though by no means fully established, is clearer, and a number of paths that previously appeared possible can now be rejected (e.g. Day 1973). Instead of a constant rate of evolution, the fossil record shows mosaicism, with some characters of the hominid repertoire evolving rapidly at some periods, other characters at others. Secondly, there have emerged enquiries into rates of molecular evolution (Sarich and Wilson 1966, 1967; Wilson and Sarich 1969; Goodman 1971; Langley and Fitch 1973), in which comparison of the aminoacid sequences in homologous proteins from different species give estimates of the number of aminoacid substitutions per unit time. Again, primate studies suggest that these rates are not constant. Dr. Ashton's contention that appreciable amounts of change can occur during evolutionarily short periods appears well validated.

In outlining modern developments in the study of heritable variation in human populations, and in indicating the selective significance of certain genetic combinations, the earlier contributions to the present symposium have focused attention on the now widely accepted thesis, that man's evolutionary development has resulted from the interplay of genetic and selective factors broadly similar to those which operate in other mammalian groups. There are, however, acting upon human populations certain unique selective forces which derive from man's capacity for abstract thought and his consequent ability to transmit tradition and culture by mechanisms dissociated from the genetic system (Huxley, 1958).

These psychological factors have resulted from the increase in size and complexity that has occurred during the evolution of the human brain. This, in turn, has been associated wtih marked modifications in the cranial and facial skeleton, while concurrently there have arisen in the post-cranial skeleton numerous distinctive features consequent upon man's assumption of an upright posture. The purpose of this paper is to examine from evidence provided by the fossil record and by populations of living Primates whether or not the emergence of man's unique morphological features has been associated with any exceptionally rapid phases of evolutionary change.

The most effective measure of evolutionary change in natural populations is the degree of divergence between their genetic systems, but this can only be assessed if the populations can be subjected to breeding tests. In the case of the Primates, the most that can be attempted is an estimation of the divergence between phenotypic characters. When the analysis is extended to the fossil remains on which detailed knowledge of the evolutionary history of the group is based, study is further restricted to morphological features. Such features can, in many cases, best be defined by measurements, and from these it is possible to derive some estimate of the overall divergence between populations. In practice, however, it is difficult to provide by means of measurements a complete description of an object of relatively simple shape—e.g., a molar tooth—and the number of dimensions necessary to give an idea of anything more than the overall proportions of one as irregular as a skull is prohibitively big. Again, it is never possible to examine more than a small sample of individuals from any one population, and in the case of fossil groups, it is not unusual for the available samples to be too small to give an adequate estimate of the means and variances of the populations from which they were drawn. Finally, although modern statistical theory has provided a basis for the comparison of constellations of interrelated measurements (e.g., see Ashton, Healy and Lipton, 1957) giving to each a weight dependent upon its contribution to the overall description, the methods are intensely laborious even when the overall numbers of measurements are limited and when electronic computing apparatus is available.

The assessment, therefore, of overall degrees of anatomical

divergence and their summary in the Linnean system of classification is still, to a large extent, based on subjective judgement. Even so, provided that the overall differences between contemporaneous genera are similar, some indication of the degree of evolutionary diversity is given by the numbers of such groups originating during any single geological epoch. If the samples of fossil populations of different geological ages are adequate to enable lineages to be deduced, and if genera in different strata are separated by similar amounts of morphological difference, an estimate of rates of change can be derived from the length of time that each genus survives (Simpson, 1953).

In certain mammalian orders (e.g., ungulates) the fossil evidence is sufficient to enable these methods to be usefully applied, but in the case of the Primates, the fossil remains are too few and fragmentary to form a basis for anything more than the most general deductions either about relationships between the main subdivisions of the order or about the rate of morphological change within each.

LIVING AND FOSSIL PRIMATES

The Primates fall naturally into two main sections: first the Prosimii which contains the lemurs and tarsiers, the most primitive members of the order, and secondly the Anthropoidea which comprises on the one hand the monkeys of the New and Old Worlds and, on the other, apes and man (Simpson, 1945).

Prosimians first appeared in the Palaeocene along with the earliest representatives of other mammalian orders, and during the succeeding Eocene the group produced numerous and varied species. Some—e.g., the Adapidae—were not dissimilar from the modern lemurs, while others—e.g., the Anaptomorphidae—inclined in their overall features towards the present-day tarsier (Simpson, 1940, 1945). In this early radiation, many genera differed from both the living lemurs and tarsiers and whereas certain groups did not reveal any marked structural peculiarities others showed a number of distinctive traits which have been variously interpreted as adaptations to different ways of life (Barth, 1951). Thus, some groups developed big incisor teeth. In others, there were elongated antero-posterior ridges on the molar teeth, while in a third group, the premolars developed into shear-like structures.

With the exception of isolated genera that persisted into the

Lower Oligocene, and of a single genus of fossil galago from the Lower Miocene of East Africa (Le Gros Clark and Thomas, 1952) nothing is known of the subsequent development of the lemurs and tarsiers until the Pleistocene, when a number of quite unusual lemurs evolved in Madagascar (Standing, 1908). One species comprised individuals that were exceptionally big. Another species developed a number of cranial features similar to those now seen in the monkeys, while in a third, there were peculiarities in the architecture of the external nares which may indicate that the animal lived in the water.

The modern lemurs and the closely-related loris form a variable group ranging from the common lemur which is, in effect, a simple tree-living mammal, through types such as the bush baby in which an elongation of the tarsal bones enables the animal to move rapidly by hopping, to such aberrant forms as the aye-aye in which the finger and toe nails characteristic of other Primates have been replaced by claws, and in which the incisor teeth resemble those of rodents (see, for example, Forbes, 1894; Elliot, 1912; Hooton, 1942).

Most of the early tarsiers died out towards the end of the Eocene, and no fossil remains have been recovered from deposits later than the Lower Oligocene. The group is now represented only by the spectral tarsier from the East Indies and Philippines.

The first fossil monkeys are represented by a small number of fragmentary jaws and teeth from the Lower Oligocene (Schlosser, 1911), the usual assumption being that their ancestors were one of the more generalized groups of Eocene prosimians. The subsequent fossil record of the monkeys is almost as fragmentary as that of the prosimians, but it appears that by the Lower Pliocene a number of groups (e.g., *Mesopithecus* and *Dolichopithecus*) closely resembling modern families had differentiated.

Present-day monkeys consist of two main groups, one living in Central and South America and the other extending throughout much of Africa and part of Asia. Both groups vary widely. That in the New World ranges from tiny and presumably primitive types such as the tamarins and marmosets to specialized groups such as the howler monkeys. That in the Old World includes numerous varieties extending from the small tree-living talapoin monkey to the big ground-living baboons (Forbes, 1894; Elliot, 1912; Hooton, 1942).

Fossil apes first appeared in the Lower Oligocene alongside the early fossil monkeys. They were small animals, and so far as can be inferred from sparse and fragmentary remains, were possibly related to the present-day gibbon and siamang. Other fossil lesser apes are known from later deposits, extending from the Lower Miocene to the Pliocene. Of these *Limnopithecus* from the Lower Miocene of East Africa is probably the best known and remains of its limb bones have suggested that these early types of ape may not have adopted the method of locomotion (brachiation) that characterizes the living members of the group (Le Gros Clark and Thomas, 1951).

The earliest great apes (*Proconsul*) appeared in East Africa during the Lower Miocene (Le Gros Clark and Leakey, 1951). Later types (e.g., *Dryopithecus,* (Gregory, 1920); *Ramapithecus* (Gregory, Hellman and Lewis, 1938)) have been recovered from deposits extending to the Upper Pliocene, but most species are known from little more than fragmentary remains of jaws and teeth.

The present-day great apes comprise the chimpanzee, gorilla and orang-utan, all of which are characterized by their peculiar method of locomotion (brachiation) with which are correlated numerous distinctive anatomical features (e.g., the excessive length of the arms). The maximum size of the brain is about 700 ccs (Ashton and Spence, 1958)—a figure some 300 ccs below the minimum size compatible with social behaviour of the type that characterizes human groups today.

Man in a clearly-defined form does not appear until the Middle Pleistocene and is represented, in the first place, by archaic hominids from Java and China now collectively referred to the genus *Pithecanthropus* (Von Koenigswald and Weidenreich, 1939), together with an apparently similar form from North Africa— *Atlanthropus mauritanicus* (Weinert, 1951). These are known principally from the remains of skulls and teeth, and differ from *Homo sapiens* in the relatively small size of the brain, in the protruding facial skeleton and heavy brow ridges, in the lack of a chin and in the large size of the cheek teeth (Weidenreich, 1937, 1943). Such remains of the post-cranial skeleton as have been discovered, provide no evidence to indicate that the posture and gait of these hominids differed from that characteristic of modern man (Weiden-reich, 1941).

Another group of archaic men, usually referred to the species *Homo neanderthalensis,* occupied parts of Western Europe during the last ice age and interglacial period (Howell, 1951). Similar types also existed in Java, South Africa and the Middle East. These groups differed anatomically from *Homo sapiens* but resembled in many of their features the earlier groups from the Far East. They differed, however, from these latter, in that the brain was quite as big as in modern man (Morant, 1927; Von Bonin, 1934). The post-cranial skeleton of this hominid is relatively well-known, and modern studies (Straus and Cave, 1957) have shown that contrary to earlier views (e.g., see Boule, 1923) there is no reason to suppose that Neanderthal man did not walk erect.

The earliest remains that can with reasonable certainty be assigned to *Homo sapiens* date only from the early phases of the last glaciation—i.e., from between 30,000 and 50,000 years ago (Oakley, 1956). However, fragmentary evidence as, for example, provided by the skulls from Swanscombe (Morant, 1938) and Fontechevade (Vallois, 1949), suggests that *Homo sapiens* may date back as far as the Middle Pleistocene.

The positions of two fossil groups—the genus *Oreopithecus* and the South African Australopithecinae—in relation to the broad outline of Primate evolution are still disputed. *Oreopithecus* is now well known from specimens recovered from deposits, possibly of early Pliocene age, in Tuscany, and although originally regarded as an Old World Monkey (Gervais, 1872) has been claimed in more recent studies to belong to the zoological group formed by the apes and man. It has been suggested that this form may even have been a hominid (Hurzeler, 1958).

The Australopithecinae are now known by abundant skeletal remains from deposits whose lower limit is given as the "second half of the early Pleistocene" (Oakley, 1955), and comprise individuals which, in the overall proportions of the brain case and facial skeleton, were far more like the living great apes than living man (Zuckerman, 1950, 1952). Certain of their features, e.g., the proportions of the canine teeth, did, however, differ from those in the living great apes, and apparent similarities between parts of their post-cranial skeleton and the corresponding regions in modern man have led to the view that in posture and gait these creatures approached *Homo* (e.g., see Le Gros Clark, 1949). On this basis, some scholars have submitted that anatomically, although not geologically, these creatures represented an ancestral stage in human evolution, and should, therefore, be included in the family Hominidae (e.g., see Le Gros Clark, 1955).

It would, in this context, be inappropriate to summarize recent discussions about the anatomical significance of the features of these fossils, but it is pertinent to mention that some of the supposed deviations from the living great apes—for example, in the size of the brain (Broom, Robinson and Schepers, 1950)—have not been substantiated by more detailed analyses (Ashton, 1950; Ashton and Spence, 1958), and the possibility cannot be excluded that other supposed differences—for example, in the proportions of the iliac blade—even if established would not necessarily have the functional significance that has been attributed to them.

Whatever fascination may attach to further anatomical studies of this group, their final allocation to either the Pongidae or Hominidae would appear to affect but little the main facts that emerge from an examination of the overall distribution of fossil Primates. Compared with many mammalian orders, the group is badly represented in the fossil record, and although most fossils can be assigned with reasonable certainty to one or other of the major subdivisions of the order, in no cases are there available series of fossils closely related to each other in successive geological strata and showing series of anatomical gradations linking two major groups. This is especially true in the case of the apes and man, and notwithstanding the numerous speculations that have been advanced little is known about the anatomical form of man's immediate ancestors. The human line of descent may, in fact, have originated at practically any point between the Lower Oligocene and the Pliocene, morphological change taking place at a completely indeterminate rate.

RATE OF CHANGE IN POPULATIONS OF LIVING

So far as can be ascertained, the only available information bearing directly on the question of rate of change in Primate populations relates to a colony of green monkeys (*Cercopithecus aethiops sabaeus*) which has been isolated for a known period of time on the West Indian Island of St. Kitts.

The green monkey has a wide distribution in West Africa extending from Senegambia to the Niger and south-eastwards to the Congo. For many hundreds of years, these animals have been exported as pets, and in the early days of the development of the sugar trade in the West Indies, green monkeys were often transported across the Atlantic along with the slaves that were needed to run the sugar plantations. It was in this way that monkeys reached St. Kitts during the early part of the seventeenth century. At this time the island was being colonized jointly by the French and British, and it appears that numbers of green monkeys escaped during the frequent quarrels which, at that time, occurred between the British and French settlers. The animals soon established free-ranging colonies and by the 1680's had become so numerous that they were officially declared vermin and a bounty was offered for every one that was killed.

Descriptions provided by travellers in the early eighteenth century, make it clear that the St. Kitts monkey was an Old World species, and not an indigenous New World variety, while specimens that were sent during the 1860's to the London Zoological Gardens, left no doubt that the animals were, in fact, the common West African green monkey. The green monkey would appear, therefore, to have been established on St. Kitts for a period of some 300 years, or if a generation is taken as being from three to four years, for between 75 and 100 generations.

Some fifteen years ago, a group of 95 skulls of the St. Kitts green monkey was sent to the Royal College of Surgeons in London, and a preliminary study (Colyer, 1948) showed that certain abnormalities, such as the presence of extra teeth, together with variations in the roots of the third molars and in the position of individual teeth, occurred more frequently in the island population than in the present-day African descendants of the parental stock.

A second group of studies of the skulls and teeth of this collection of St. Kitts green monkeys brought to light the fact that significant differences now exist between corresponding cranial and dental dimensions of the island and mainland populations. The skulls and teeth of the island monkeys tend to be bigger than those of the present-day African green monkey, while the variability of its cranial and dental dimensions is now significantly less (Ashton and Zuckerman, 1950, 1951a). Variation between corresponding structures on the left and right sides of the mid-line is, on the other hand, now significantly greater in the skulls and teeth of the St. Kitts green monkey than in the African variety—in other words, the St. Kitts monkeys are now less symmetrical than the present-day African ones (Ashton and Zuckerman, 1951b).

The degree of difference between the teeth of the St. Kitts and African green monkeys is similar to that which now separates the African green monkey from certain other groups of cercopitheques which, on the basis of coat coloration and geographical distribution are usually given distinct subspecific or specific status. A difference of this size arising in the course of 300 years represents a rate of change between 5,000 and 10,000 times that which has characterized the evolution of horse's teeth during the past 10 million or so years (Ashton and Zuckerman, 1951c).

In the absence of data derived from direct breeding experiments, the possibility cannot be excluded that the observed morphological differences between the St. Kitts and African green monkeys are due to the action of a different environment upon an unaltered genetic constitution. On the other hand, an indication that the changes may have a genetic basis is given by the fact that the total pattern of change in the island monkey is in many ways analogous to that which has been recorded in certain laboratory experiments as a result of the action of selection upon systems of multiple genes (see, for example, Mather and Harrison, 1949; Sismanidis, 1942).

Even if this view is correct, it is doubtful whether the high rate of change that has occurred during the first 300 years of isolation would be maintained indefinitely. The island stock is continually losing free variability and it would seem reasonable to suppose that continued selective pressure on the St. Kitts monkeys would relatively rapidly result in its extinction. The observations do, however, show that in the higher Primates, measurable amounts of change can occur during periods which, by geological standards, are quite insignificant.

References

Ashton, E. H. (1950) The endorcranial capacities of the Australopithecinae. *Proc. Zool. Soc. Lond.* **120**, 715.

Ashton, E. H. (1957) The use of quantitative methods in the study of primate evolution. *Scientia, Bologna* **92**, 232.

ASHTON, E. H., HEALY, M. J. R. and LIPTON, S. (1957) The descriptive use of discriminant functions in physical anthropology. *Proc. Roy. Soc.* B **146**, 552.

ASHTON, E. H. and SPENCE, T. F. (1958) Age changes in the cranial capacity and foramen magnum of hominoids. *Proc. Zool. Soc. Lond.* **130**, 169.

ASHTON, E. H. and ZUCKERMAN, S. (1950) The influence of geographic isolation on the skull of the green monkey (*Cercopithecus aethiops sabaeus*) i A comparison between the teeth of the St. Kitts and the African green monkey. *Proc. Roy. Soc.* B **137**, 212.

ASHTON, E. H. and ZUCKERMAN, S. (1951a) The influence of geographic isolation on the skull of the green monkey (*Cercopithecus aethiops sabaeus*) ii The cranial dimensions of the St. Kitts and the African green monkey. *Proc. Roy. Soc.* B **138**, 204.

ASHTON, E. H. and ZUCKERMAN, S. (1951b) The influence of geographic isolation on the skull of the green monkey (*Cercopithecus aethiops sabaeus*) iii The developmental stability of the skulls and teeth of the St. Kitts and African green monkey. *Proc. Roy. Soc.* B **138**, 312.

ASHTON, E. H. and ZUCKERMAN, S. (1951c) The influence of geographic isolation on the skull of the green monkey (*Cercopithecus aethiops sabaeus*) iv The degree and speed of dental differentiation in the St. Kitts green monkey. *Proc. Roy. Soc.* B **138**, 354.

BARTH, F. (1951) On the relationships of early Primates. *Am. J. Phys. Anthrop.* (n.s.) **8**, 139.

BONIN, G. VON (1934) On the size of man's brain as indicated by skull capacity. *J. Comp. Neurol.* **59**, 1.

BOULE, M. (1923) *Fossil Men—Elements of Human Palaeontology.* (English translation of 2nd French edition). Gurney & Jackson, London; Oliver & Boyd, Edinburgh.

BROOM, R., ROBINSON, J. T. and SCHEPERS, G. W. H. (1950) Sterkfontein ape-man, Plesianthropus. *Trans. Mus. Mem.* **4**, 1.

CLARK, W. E. LE GROS (1949) New palaeontological evidence bearing on the evolution of the hominoidea. *Quart. J. Geol. Soc. Lond.* **105**, 225.

CLARK, W. E. LE GROS (1955) *The Fossil Evidence for Human Evolution— An Introduction to the Study of Palaeoanthropology.* Chicago University Press.

CLARK, W. E. LE GROS and LEAKEY, L. S. B. (1951) *The Miocene Hominoidea of East Africa.* (Fossil mammals of Africa No. 1), British Museum, London.

CLARK, W. E. LE GROS and THOMAS, D. P. (1951) *Associated Jaws and Limb Bones of Limnopithecus Macinnesi.* (Fossil mammals of Africa No. 3), British Museum, London.

CLARK, W. E. LE GROS and THOMAS, D. P. (1952) *The Miocene Lemuroids of East Africa.* (Fossil mammals of Africa No. 5), British Museum, London.

COLYER, J. F. (1948) Variations of the teeth of the green monkey in St. Kitts. *Proc. Roy. Soc. Med.* **41**, 845.

ELLIOT, D. G. (1912) A review of the Primates. (3 vols.) *Monog. Amer. Mus. Nat. Hist.* **1**, 1.

FORBES, H. O. (1894) *A Handbook to the Primates.* (2 vols.) Allen, London.

GERVAIS, P. (1872) Sur un singe fossile, d'espèce non encore décrite, qui a été découvert au Monte-Bamboli (Italie). *C. R. Acad. Sci., Paris* **74**, 1217.

GREGORY, W. K. (1920) The origin and evolution of the human dentition. *IV* The dentition of the higher Primates and their relationships with man. *J. dent. Res.* **2**, 607.

GREGORY, W. K., HELLMAN, M. and LEWIS, G. E. (1938) Fossil anthropoids of the Yale-Cambridge India expedition of 1935. *Publ. Carnegie Instn.* **495**, 1.

HOOTON, E. (1942) *Man's Poor Relations*. Doubleday, New York.

HOWELL, F. C. (1951) The place of neanderthal man in human evolution. *Amer. J. Phys. Anthop.* (n.s.) **9**, 379.

HÜRZELER, J. (1958) *Oreopithecus bambolii* Gervais—a preliminary report. *Verh. naturf. Ges. Basel* **69**, 1.

HUXLEY, J. S. (1958) Man's place in nature. *Sunday Times*, 13 July.

KOENIGSWALD, G. H. R. VON and WEIDENREICH, F. (1939) The relationship between *Pithecanthropus* and *Sinanthropus*. *Nature, Lond.* **144**, 926.

MATHER, K., and HARRISON, B. J. (1949) The manifold effect of selection. *Heredity* **3**, 1 and 131.

MORANT, G. M. (1927) Studies of palaeolithic man *ii* A biometric study of neanderthaloid skulls and of their relationships to modern racial types. *Ann. Eugen., Lond.* **2**, 318.

MORANT, G. M. (1938) The form of the Swanscombe skull. *J. Roy. Anthrop. Inst.* **68**, 67.

OAKLEY, K. P. (1955) Dating the australopithecines. *Proc. Pan-Afr. Congr. Prehist.* **3**, 155.

OAKLEY, K. P. (1956) Dating fossil men. *Mem. Manchr. Lit. Phil. Soc.* **98**, 1.

SCHLOSSER, M. (1911) Beiträge zur Kenntnis der Oligozänen Landsäugetiere aus dem Fayum (Ägypten). *Beitr. Paläont. Geol. Öst-Ung.* **24**, 51.

SIMPSON, G. G. (1940) Studies on the earliest Primates. *Bull. Am. Mus. Nat. Hist.* **77**, 185.

SIMPSON, G. G. (1945) The principles of classification and a classification of mammals. *Bull. Am. Mus. Nat. Hist.* **85**, 1.

SIMPSON, G. G. (1953) *The Major Features of Evolution*. Columbia University Press, New York.

SISMANIDIS, A. (1942) Selection for an almost invariable character in *Drosophila*. *J. Genet.* **44**, 204.

STANDING, H. F. (1908) On recently discovered subfossil Primates from Madagascar. *Trans. Zool. Soc. Lond.* **18**, 59.

STRAUS, W. L. JR. and CAVE, A. J. E. (1957) Pathology and the posture of neanderthal man. *Quart. Rev. Biol.* **32**, 348.

WEIDENREICH, F. (1937) The dentition of *Sinanthropus pekinensis:* a comparative odontography of the hominids. *Palaeont. Sinica.* **101**, 1.

WEIDENREICH, F. (1941) The extremity bones of *Sinanthropus pekinensis*. *Palaeont. Sinica* **116**, 1.

WEIDENREICH, F. (1943) The skull of *Sinanthropus pekinensis:* a comparative study on a primitive hominid skull. *Palaeont. Sinica* **127**, 1.

WEINERT, H. (1951) Über die neuen Vor- und Frühmenschenfunde aus Africa, Java, China and Frankreich. *Z. Morph. Anthr.* **42**, 113.

VALLOIS, H. V. (1949) The Fontéchevadè fossil men. *Am. J. Phys. Anthrop.* (n.s.) **7**, 339.

ZUCKERMAN, S. (1950) Taxonomy and human evolution. *Biol. Rev.* **25**, 435.

ZUCKERMAN, S. (1952) An ape or the ape. *J. Roy. Anthrop. Inst.* **81**, 57.

BLOOD GROUPS

A. E. MOURANT

In the thirteen years since this paper was written, the story of the blood groups has been one of generally increasing complexity. The first X-linked blood group system (Xg) was discovered in 1962, the Dombrock in 1965, and Auberger in 1961; the Ko and Zw platelet groups in 1962 and 1963; the serum groups (Lp and Xm) in 1963. But it is in the increased complexity of the Gm and Inv groups that the variation in human populations reaches its most intricate (Race and Sanger 1968; Grubb 1970). Secondly, there has been the great development of the HL-A histocompatibility types which show a complexity of variants and population specificity similar to the Gm. The third main field of investigation has been the intensive search for associations with diseases and fertility in an attempt to establish the selective value of the blood group systems (e.g. Cohen 1970; Reed 1968; Vogel 1973). A particularly interesting development has been the recent success in biochemical differentiation of blood groups, with the evidence that the transferases of the A_1 and A_2 blood groups are different enzymes (Schachter et al. 1973).

THE first systematic application to anthropology of the ABO blood groups, discovered in 1900, was made by Professor and Mrs. Hirszfeld in 1918 (Hirszfeld and Hirszfeld, 1918–19, 1919), when they determined the groups of many hundreds of persons from many different parts of the world. Since then there have been alternating periods of advance in basic knowledge of blood groups, and of the application of this knowledge to population studies, the results of which, in some cases, have in turn led to basic theoretical advances.

At the present time some eleven genetically independent blood group systems are known, and studies have been carried out of the distribution of the antigens of all these systems in a variety of populations. While, however, the results of ABO grouping of over 7,000,000 people have been published, drawn from every part of the world, the numbers tested for the other systems are considerably less, varying from some hundreds of thousands for the Rh groups to a few hundred for the Js groups, the latest to be discovered.

In general, even the very latest large-scale studies, insofar as results have been published, have nearly all perforce been carried out and reported in terms of our theoretical knowledge of at least two or three years ago, as described, for instance, in the third edition of Race and Sanger (1958), fully up-to-date as that was at the time of publication. This limitation has been due to the lack of adequate quantities of certain newly discovered critical testing sera, and to the fact that few anthropologists, and indeed few serologists and geneticists, have been able so far to assimilate all the results of the latest very complex but fundamentally important studies on some of the blood group systems, especially the Rh system. Indeed, the accessions of new knowledge of a theoretically disturbing nature are occurring at such a rapid rate that perhaps no single worker in any of the disciplines just mentioned has come near to establishing a clear picture of the whole situation in his own mind, let alone communicating it to other workers.

Until recently, most workers have thought and written of the Rh system in terms of the genetical and notational scheme orginally devised by Sir Ronald Fisher (Race, 1944) and developed by him in conjunction with Dr. R. R. Race (Fisher and Race, 1946). The antigens of the Rh system are considered, according to the hypothesis underlying this scheme, to be the effects of genes at three very closely linked loci on a single chromosome. The genotype of a given individual, and the phenotype, or combination of observed serological reactions of the red cells, must, of course, be referred to the combined effects of two of these triplets of genes, each occurring on one of a pair of corresponding chromosomes. The loci are known as C, D and E, and are thought to be in the linear order D C E. At each locus one of a set of allelomorphic genes occurs, D or d, C or c, E or e (disregarding a number of rare allelomorphs which are known at each of the three loci). Thus the genotype of a given individual consists of two sets of three genes, such as DCe and dce, which are represented in a genotype symbol such as, in the instance cited, DCe/dce. Each gene present in the genotype is represented by a corresponding antigen in the phenotype, except that no antigen has hitherto been detected corresponding to the gene d. Thus a DCe/dce person will carry on his or her red cells the antigens D, C, c and e, each of them detectable by means of an agglutination test with a serum containing the corresponding antibody (but apparently nothing detectable corresponding to the d gene).

In the original form of the scheme, the particular distribution of a given set of genes between the two chromosomes was regarded as not affecting the phenotype. Thus, for instance, the genotypes DCe/dce and Dce/dCe were considered indistinguishable without family studies.

Gradually, however, in recent years, a body of facts has emerged which cannot be explained in terms of the simple hypothesis which I have just outlined. The discoveries involved have mostly been of new distinct antigens, each inherited as part of the closely linked Rh system, but not explicable as a simple product of an allelic gene at any one of the three established loci.

Many workers, especially Race and Sanger, think that all the facts, if not already explicable in terms of simple modifications of the system of Fisher and Race, will ultimately prove to be explicable in this way. Other workers disagree to a varying extent with this view. The differences of opinion which exist depend partly upon personal observations of the reactions of unique testing reagents prepared from the sera of individual human beings, who have suffered unusual kinds of immunization, and partly upon differences in the interpretation of new facts which are, themselves, generally accepted. In general, given time, apparent factual differences are sorted out, but at any given moment some such differences usually persist. Differences of interpretation are not so easily reconciled. In a short paper such as this it is impossible to set out the multiplicity of relevant published observations, let alone the unpublished ones known to me, divorced from any theoretical explanations. All that can be done is to introduce some of the new facts in terms of that theoretical framework which appears to give the most likely explanations of them, while admitting that some competent workers disagree with these interpretations. No originality is claimed for these explanations which are, in their essentials, those put forward by Race and Sanger (Race, 1960).

One of the simplest of these sets of observations is that relating to the antigen known as f. This was at first regarded as a product of a gene at a fourth locus closely linked to the other Rh loci. It has been noticed, however, that nearly always, and perhaps invariably, f is a product of an Rh chromosome which carries the genes c and e. It should be noted that in order for f to be present the c and e must be on the same chromosome and not on opposite chromosomes in the genotype. Now, therefore, the antigen f is regarded as a product of the genes c and e acting together. In terms of modern genetics, since c and e do not act together in this way when on different chromosomes, these genes are regarded as noncomplementary and should, if we may draw an analogy with microbiological genetics, be regarded as forming part of a single *cistron* or linearly arranged closely linked group of sub-genes each behaving as a separate entity with respect to mutation and crossing-over. So regarded, the observations thus give considerable support to the view that the Rh genes, at any rate those of the C and E classes, form a single closely linked group.

A more complex group of facts, of considerable anthropological importance, refers mainly to observations on Africans. An antigen V (De Natale *et al.*, 1955), inherited as part of the Rh complex, is present in about 40 per cent of Negroes, but occurs only very exceptionally in Europeans. When present, it is always associated with dce, D^{u}ce or Dce, but unlike f it does not by any means invariably accompany these combinations.

Another observation, at first apparently unrelated to V, and first clearly reported by Sturgeon (1960) is the fact that the chromosome combination usually diagnosed as dCe differs in Europeans and Negroes. In Europeans, with exceedingly rare exceptions, it produces an antigen (C) which reacts with all anti-C sera. In Negroes, however, the corresponding C antigen reacts variably with sera hitherto accepted as all having anti-C specificity, and fails to react with some of them. The so-called anti-C sera with which it fails to react prove to be of a kind first distinguished and described by Rosenfield and Haber (1958) who showed that some "anti-C" sera from Rh-positive individuals have in fact a specificity anti-Ce, reacting only with an antigen produced jointly by the C and e genes when these are present at the same chromosome. This did not, however, explain the negative reactions with Negro dCe which appeared to contain both C and e on one chromosome.

A solution was found by Sanger *et al.* (1960) when they examined a serum containing a peculiar and hitherto unknown type of antibody. This antibody was found to react both with bloods containing the V antigen and with those containing "Negro dCe". The antibody in this serum was a single one, not separable into components of different specificities by any known technique. The authors explained this peculiar combination of reactions by postulating that Negroes commonly, but Europeans very rarely, have a special allele of E and e which they call e^s. The antigen V is a joint product of c and e^s found in the chromosome combinations dce^s, D^uce^s and Dce^s. The Negro chromosome hitherto described as dCe is in fact dCe^s and hence cannot give rise to a product Ce reacting with the specific anti-Ce.

As already mentioned, the modern theory of the structure of chromosomes in general is entirely compatible with the hypothesis that the Rh blood groups are determined by a series of three or more closely linked genes. It is a fact that even a gene which, to the formal geneticist, appears perfectly simple is a chain composed of a very large number of deoxyribonucleic acid (DNA) units. The general theory gives far more than enough scope to allow for all the minute variations which have been described for the Rh system. The immediate problem is not whether the theory is adequate, but whether separate chromosome sub-segments within the major Rh segment are responsible for the C, and D and the E allelomorphs respectively, and, if so, whether we can ascertain their linear order. While we cannot at present be completely certain of the existence of such discrete sub-segments, or of their order, the evidence is highly suggestive, and the supposition that they exist, and are in the order DCE, has served for many years now as a very powerful working hypothesis.

The study of the blood groups of human populations in general is becoming less and less a series of accidental off-shoots of blood transfusion work, and more and more a planned and integrated investigation. Indeed not only are blood group studies being planned so as to include data on as many blood group systems as possible, but they are being planned in conjunction with the study of other genetical systems, such as those determining the haemoglobins and plasma proteins like the haptoglobins. We may perhaps hope that genetical population studies will one day include also an examination of leucocyte and platelet antigens, but the technical problems involved are considerable and there are no signs that the day is near.

It is fortunate that suitable tests on a single relatively small sample of blood can already give information regarding the genes not only of the eleven major blood group systems, but of numerous other genetical systems as well.

Mention must, however, be made of two genetical systems with biochemical manifestations detectable other than in blood specimens, namely by the ability or inability to detect a bitter taste in phenylthiocarbamide, and by the secretion or non-secretion in the saliva of the specific ABO (or strictly, ABH) blood group substances. They are mentioned here rather to urge the need for supplementing the present inadequate population data than to record any recent advances, other than in the technique of taste testing which, especially in the hands of Dr. H. Kalmus (1958) and his colleagues, has in recent years become much more precise than hitherto.

While integrated population studies, largely based on very full series of tests on blood samples, are being performed on a quite extensive scale on selected populations in many parts of the world, the total numbers of individuals tested is, nevertheless, limited by the time and effort involved in carrying out the tests, and in the case of some blood group systems by the rarity of the reagents.

Because these limitations to the numbers that can be tested apply to a much lesser degree to the ABO and Rh systems, even more extensive surveys are being done of the distribution of the blood groups of these systems. The material is being collected from two main sources—from transfusion services generally, most of which serve large centres of population, especially in industrial countries; and from anthropologically directed surveys in areas where facilities for more comprehensive genetical surveys are limited or lacking.

Thus practically every country in the world has been covered more or less completely by ABO surveys, but even in the most intensively studied areas, such as some of the countries of Western Europe, there is a need for a more detailed analysis of ABO distribution than at present exists.

While a survey embodying only the ABO groups, or only these

and the Rh groups, is of great importance for a preliminary genetical classification of the population concerned, it is quite inadequate for any final classification, or assessment of relationships between populations, since resemblances in the frequencies of the genes of only one or two systems may be accidental and not due to community of origin. Two or more populations can only be regarded with confidence as closely related if they have been shown to resemble one another with regard to gene frequencies of several blood-group or other genetical systems. Gradually the peoples of Europe, and selected populations of special interest elsewhere, are being examined for the distribution of the genes of a great many systems, and from time to time comprehensive reviews of the situation in one or several countries are made.

One of the most thorough blood group surveys of a single country yet carried out is that done by Beckman (1959) in Sweden. It is of particular interest since he has shown that the frequencies of the genes in the ABO, MN, P and Rh systems vary in concordance with one another and thus that the populations of the various parts of Sweden can be regarded to a fair degree of approximation as made up of varying proportions of only three genetical components or ancestral strains, the West European, the Eastern, and the Lapp.

The Western strain, which is the one mainly of interest to us at the moment, is characterized by high, or relatively high, frequencies of O, N, d (Rh-negative) and P_1, and is low in B. The Eastern strain has relatively high A_1 and B, and also, as regards its "East Baltic" component, high M. The Lapp strain has high frequencies of A_2, D, N and P_2. Outside Sweden the Lapp strain is likely to be confined to northern Norway and Finland, and north-western Russia, but if the peoples of the rest of Europe are as simple in their genetical origins as those of Sweden, one would expect to find the other two strains, together with only a limited number of others, emerging from a genetical analysis of the populations of other countries. That they have, in general, not yet so emerged must be attributed mainly to lack of investigation. However, in Ireland a western component with a very high O frequency does emerge very clearly. It differs, however, from that deduced from the Swedish data in having a rather high frequency of B and a lower frequency of d. In Eire, as a whole there is in fact, contrary to what is found in Sweden, a negative correlation between O and d, though the opposite is true in the Dublin area considered by itself (Dawson and Hackett, 1958).

Something closely resembling the Irish high-O strain is found in Scotland, North Wales and Iceland. The Basques also have a high O frequency but in several respects they are more like the Swedish high-O strain.

Clearly what is now needed is an analysis of the peoples of other European countries as detailed as that now available for Sweden, especially with regard to the blood group systems other than the ABO. Unfortunately, we can expect the necessary data for most countries to be obtained only very slowly. In Sweden they were derived from the results of medico-legal paternity testing. The only other country in Europe where this is done on a similar scale to Sweden is Denmark, which is a small and probably highly homogeneous country.

In most other countries we can expect to acquire the necessary data only slowly as a result of investigations related to blood transfusion, of a limited volume of medico-legal testing, and of work carried out primarily for research purposes.

It should be added that, in his analysis of the population of Sweden, Dr. Beckman also took into account the very extensive anthropometric data available, and showed that they led to the same conclusions as did the blood groups.

While investigation of the distribution of numerous factors is the ideal, it is clear that the numbers of individuals who can be investigated in this way cannot exceed a few thousand in any one country. On the other hand, data on the ABO groups of many hundreds of thousands of individuals can be obtained from the records of transfusion services in many countries, and information of this kind is leading to highly detailed local classification, even if the broader connections of the classes so defined cannot be ascertained without the help of the other blood group systems.

Thus in Europe we are gradually achieving a detailed genetical analysis of populations, most of which are stable and have no very marked internal barriers to interbreeding. In the other continents we may expect that a similar type of detailed analysis will ultimately be applied to other relatively stable populations, such as those of

China and Japan. Elsewhere the problems facing population geneticists are very different.

In India the caste system presents a particular problem. Whatever may be thought of the present social value of the caste system, there is no question as to its importance as a subject for genetical investigation, and if for political or other reasons the castes are gradually to become merged, it is all the more important that the nature of the genetical differences between them should be investigated immediately. The importance of this is widely appreciated by Indian anthropologists, and numerous relatively small investigations of ABO groups are being carried out, but only a very few centres which have the necessary resources are going beyond this into an analysis based on several blood group systems.

It must be admitted that no very clear picture has emerged as to the genetical nature of caste differences. It is clear that underlying genetical differences exist, and that they may, as Sanghvi and Khanolkar (1949) have shown in Bombay, be very marked. Much more marked, however, are the differences between the settled peoples of India and the aboriginal tribes. The former fall broadly into the Mediterranean picture, while some of the latter suggest affinities rather with south-east Asia or even Australia.

In the Near East we find numerous "ethnic" groups living in close proximity but not intermarrying. Here the heterogeneity appears likely to maintain its stability longer than that of the Indian peoples, but the problems of origin are just as interesting and perhaps easier to solve. Much work is in progress in this area but little has been published. In a few years' time we may have a much clearer picture than at present.

Of the vast areas of the world settled by recent immigration from Europe there is little to say at the moment, though genetical studies may one day prove to have much to contribute to an analysis of these people, especially when the peoples of Europe themselves are better understood.

Over most of the rest of the earth we are faced with a vast number of tribes just emerging from barbarism, and being rapidly introduced to European civilization in one or other of its forms, and losing their separate identity. Here the problem of genetical characterization is an urgent one and on the whole the challenge is being met, though not as quickly as could be wished.

Africa presents a special problem, since a considerable number of blood group antigens, and the corresponding genes, are being shown to be confined entirely, or almost so, to populations of African origin. These genes have in most cases been first found in the diluted Negro populations of America, and only subsequently in "pure" Africans. Their distribution within Africa now needs to be traced. This is being done, but much more slowly than is desirable, even in those large areas where research is at present possible, and meanwhile, in other areas, political unrest is rendering such work difficult or impossible. Outside Africa these same genes are proving of value in tracing the amount of African ancestry present, as, for instance, in the peoples of the Near East.

It may be of some interest to list the blood group characters which are peculiar, or almost peculiar, to Africans. The Rh system is of the greatest interest in this respect; it has long been known that Africans have a uniquely high frequency of the Rh chromosome Dce or R_0. They are almost alone in possessing the Rh antigen V (the most useful single marker of African ancestry, De Natale et al., 1955) and a peculiar chromosome closely related to dCe but lacking some of its reactions (Sturgeon, 1960). As I have already explained, Sanger et al. (1960) have recently explained these two almost uniquely African characteristics as different expressions of a single entity, the gene e^s.

In the MNSs systems Negroes are almost unique in possessing the Henshaw (or He) and Hunter (or Hu) antigens, both determined by genes closely linked with the MNSs complex (Chalmers et al., 1955). They are also unique in possessing S^u, the third or silent allele of S and s (Wiener et al., 1953, 1954; Greenwalt et al., 1954). Homozygotes for this gene fail to give reactions with "anti-U", a single antibody reacting in most respects like a mixture of anti-S and anti-s, as well as with anti-S and anti-s themselves.

Negroes in considerable numbers also possess a silent allele, Fy, in place of Fy^a or Fy^b of the Duffy system (Sanger et al., 1955).

The Js^a antigen (Giblett, 1959), a product of the single recognizable gene of the recently discovered Js system, is also almost entirely confined to Africans.

This genetical uniqueness of Africans is in itself an interesting problem. It would appear to result from a long genetical isolation of Africans from the rest of mankind—an isolation by no means complete, but sufficient to allow the production of mutations and their subjection to natural selection almost entirely separately from the corresponding process going on in the rest of the human race. Considerable mobility within Africa probably led to the rather thorough dispersal, throughout that part of the continent south of the Sahara, of the various genes peculiar to that region, genes which probably originated as mutations in widely different parts of the region.

In a continent, the greater part of which is almost without written history, apart from the last two or three hundred years, the blood groups have probably much to contribute to the elucidation of the story of the continent and its peoples. Already they have shown that the Bushmen and the Hottentots (Zoutendyk *et al.*, 1953, 1955), sometimes supposed to be of an entirely different race from the Negroes, are, in fact, like them, typical Africans, though with certain small peculiarities due probably to a relatively early isolation from the main Negro stock.

While no other major population group shows anything like the same degree of uniqueness as the Africans, the Diego antigen, the sole antigen known in the system of the same name, is found exclusively, or almost so, in Mongoloid peoples. By far the highest frequencies are found in the South American Indians, with the North American Indians and the Japanese next (Layrisse, 1958). The antigen is, however, almost absent among the Eskimos and apparently entirely so in the Polynesians and other Pacific Island peoples. The explanation of this remarkable distribution must almost certainly be closely related to the story of the original peopling of the American continent, but its meaning is at present very far from clear. Other essential clues for the elucidation of this story must lie in Siberia (Levin, 1959), where there has recently been a revival of blood group anthropological research, though not so far including tests for the Diego antigen.

I have recently discussed elsewhere in some detail (Mourant, 1960a, 1960b) the distribution of the blood groups in the Pacific area and in Siberia and shall, therefore, summarize the facts only very briefly now. There is a very marked difference, in blood group frequencies generally, between the peoples of eastern Asia and the aboriginal peoples of America. Of all non-American peoples (apart from the few Eskimos in Siberia) those who most closely resemble

American Indians and Eskimos in their blood group frequencies are the Polynesians, though there are some important differences. It can certainly not be taken as proved, as Heyerdahl (1950) holds, that the ancestral Polynesians came from America, but there can be little doubt that they and the aboriginal Americans have a higher proportion of recent ancestors in common than have, for instance, the Americans and the eastern Asiatics.

One respect in which American Indians and Eskimos differ from all eastern Asiatic and Pacific island peoples is in the extremely high frequency of the M gene in the former. The recent demonstration of a high M frequency in a small sample of Lamuts, a Tungus-speaking people, suggests the possibility that the Tungus peoples of north central Asia may be closely related to the pre-Columbian Americans.

As soon as it was first shown that different populations had different blood group frequencies, the problem of the origin of such differences arose. At first the favoured hypothesis was that of genetic drift, but more recently, with the development of population genetics, it has been realised that many mechanisms are probably at work, though natural selection is probably the chief one. As to both the fact of selection and its mechanism, Ford saw the situation very clearly as early as 1945 and predicted that blood group frequencies in a number of diseases would prove to differ from those in the general population, because of the greater liability of persons of one group than of another to a given disease. For a great many years workers have been investigating this problem, but the first results to be entirely convincing and to stand the test of time were those of Aird, Bentall and Roberts (1953) who demonstrated the existence of a highly significant excess of group A among sufferers from carcinoma of the stomach. Since then numerous workers have entered the field, notably Drs. Clarke and McConnell and their colleagues at Liverpool, and many diseases have been investigated. The results have been discussed from the standpoint of natural selection by Sheppard (1959) in an address to this Society, and by myself (Mourant, 1959). I therefore do not propose to discuss this subject at any length. It is, however, clear that we are now in a better position than ever before to find out what causes the differences in frequency of the blood groups, and perhaps even what causes the blood groups themselves.

Blood group studies are probably at present more valuable as a source of genetical information about human populations than studies of all other factors combined, and blood groups will surely remain for many years more important than any other single class of factors. In the past decade, however, as I have already mentioned, knowledge of genetical systems other than those of the blood groups, for which human populations are polymorphic, has multiplied very greatly, and there is no doubt that many more such systems will very soon be discovered. Among the known classes of characters to be discussed are the haemoglobins, the plasma proteins, the enzymes and their deficiencies. Genetically, and as anthropological markers, the genes involved do not differ essentially from those of the blood groups though, because of their different physiological expression, they must be subject to different kinds of mechanism of natural selection and of the establishment of stable polymorphisms. Nevertheless, many of the genetical and anthropological problems are common to the blood groups and the other classes of character, and the statistical methods of dealing with them are also closely similar. This, and the further fact, already mentioned, that most of the necessary tests can be done on a single specimen of blood from each member of the population, is having the fortunate effect of maintaining human population genetics, despite the breadth of its subject matter, as a largely unified discipline.

References

AIRD, I., BENTALL, H. H. and ROBERTS, J. A. F. (1953) Relation between cancer of stomach and ABO blood groups. *Brit. Med. J.* **1,** 799.

BECKMAN, L. (1959) A contribution to the physical anthropology and population genetics of Sweden. Doctorate thesis, Lund. *Hereditas* **45**.

CHALMERS, J. N. M., IKIN, ELIZABETH, W. and MOURANT, A. E. (1955) A study of two unusual blood-group antigens in West Africans. *Brit. Med. J.* **ii,** 175.

DAWSON, G. W. P. and HACKETT, W. E. R. (1958) A blood group survey of the county and city of Dublin. *Ann. Hum. Genet.* **22,** 97.

DE NATALE, A. *et al.* (1955) V, a "new" Rh antigen, common in Negroes, rare in white people. *J. Amer. Med. Ass.* **159,** 247.

FISHER, SIR RONALD and RACE, R. R. (1946) Rh gene frequencies in Britain. *Nature (Lond.)* **157,** 48.

FORD, E. B. (1945) Polymorphism. *Biol. Rev.* **20,** 73.

GIBLETT, ELOISE R. (1959) Js^a, a "new" red-cell antigen found in Negroes; evidence for an eleventh blood group system. *Brit. J. Haematol.* **5,** 319.

GREENWALT, T. J. *et al.* (1954) An allele of the S(s) blood group genes. *Proc. Nat. Acad. Sci., Wash.* **40,** 1126.

HEYERDAHL, T. (1950) *American Indians in the Pacific; the Theory Behind the Kon-Tiki Expedition.* Allen & Unwin, London.

HIRSZFELD, L. and HIRSZFELD, HANNA (1918–19) Essai d'application des méthodes sérologiques au problème des races. *Anthropologie, Paris* **29**, 505.

HIRSZFELD, L. and HIRSZFELD, HANNA (1919) Serological differences between the blood of different races. The result of researches on the Macedonian front. *Lancet* **ii**, 675.

KALMUS, H. (1958) Improvements in the classification of the taster genotypes. *Ann. Hum. Genet.* **22**, 222.

LAYRISSE, M. (1958) Anthropological considerations of the Diego (Di^a) antigen. *Amer. J. Phys. Anthrop.* n.s. **16**, 173.

LEVIN, M. G. (1959) Novye materialy po gruppam krovi u eskimosov i lamutov (New blood group data among Eskimos and Lamuts). *Sov. Etnogr.* No. 3, 98.

MOURANT, A. E. (1959) Human blood groups and natural selection. *Cold Spr. Harb. Symp. Quant. Biol.* **24**, 57.

MOURANT, A. E. (1960a) The relationship between the blood groups of Eskimos and American Indians and those of the peoples of eastern Asia. Paper read at Alaska Science Conference, Anchorage, Alaska.

MOURANT, A. E. (1960b) Achievements and unsolved problems of blood group anthropology. Paper read at Eighth Congress of the International Society of Blood Transfusion, Tokyo.

RACE, R. R. (1944) An "incomplete" antibody in human serum. *Nature (Lond.)* **153**, 771.

RACE, R. R. (1960) The genetics of Rh. *Proc. VIII Cong. Int. Soc. Blood Transf.* (Tokyo), 109.

RACE, R. R. and SANGER, RUTH (1958) *Blood Groups in Man,* 3rd Ed. Oxford, Blackwell.

ROSENFIELD, R. E. and HABER, GLADYS V. (1958) An Rh blood factor rh_i (Ce) and its relationship to hr (ce). *Amer. J. Hum. Genet.* **10**, 474.

SANGER, RUTH, *et al.* (1960) An Rh antibody specific for V and R'^S. *Nature (Lond.)* **186**, 171.

SANGER, RUTH, RACE, R. R. and JACK, J. A. (1955) The Duffy blood groups of New York Negroes; the phenotype Fy (a-b-), *Brit. J. Haematol.* **1**, 370.

SANGHVI, L. D. and KHANOLKAR, V. R. (1949) Data relating to seven genetical characters in six endogamous groups in Bombay. *Ann. Eugen. (Lond.)* **15**, 52.

SHEPPARD, P. M. (1959) Natural selection and some polymorphic characters in man. *Symp. Soc. Stud. Hum. Biol.* **2**, 35.

STURGEON, P. (1960) Studies on the relation of anti-rh'^N (C^N) to Rh blood factor rh_i (Ce). *J. Forens. Sci.* **5**, 287.

WIENER, A. S., UNGER, L. J. and GORDON, EVE B. (1953) Fatal hemolytic transfusion reaction caused by sensitization to a new blood factor U. *J. Amer. Ass.* **153**, 1444.

WIENER, A. S., UNGER, L. J. and COHEN, LAURA (1954) Distribution and heredity of blood factor U. *Science* **119**, 734.

ZOUTENDYK, A., KOPEĆ, ADA C. and MOURANT, A. E. (1953) The blood groups of the Bushmen. *Amer. J. Phys. Anthop.* n.s. **11**, 361.

ZOUTENDYK, A., KOPEĆ, ADA C. and MOURANT, A. E. (1955) The blood groups of the Hottentots. *Amer. J. Phys. Anthop.* n.s. **13**, 691.

ABNORMAL HAEMOGLOBIN AND ERYTHROCYTE ENZYME-DEFICIENCY TRAITS

A. C. ALLISON

Interest in these topics has continued to be lively for, on account of the demonstrably great selection differential attaching to haemoglobin S, it has proved a valuable model for understanding natural selection in man, and particularly for its illustration of the problems of methodology of studying selection. The fact that the physicochemical bases of haemoglobin variant genes were the first to be identified has produced continuing research activity, and now the aminoacid substitutions responsible for many of the haemoglobin and glucose-6-phosphate dehydrogenase variants have been established. Many new variants have been discovered and reference centres for their identification established. With the increasing complexity, the pattern of world distribution of haemoglobin variants has been sketched in greater detail (Livingstone 1973), though the outline remains very much as in this paper, and the distribution of the subtypes is emerging.

A related major field of advance, the many other red cell isoenzymes, has been very fruitful as regards the discovery of variant forms (Giblett 1969; Harris 1970), and for several the outlines of the world distribution have begun to emerge, but there is as yet very little other evidence, comparable with that available for the traits discussed by Dr. Allison, as to their selective role.

IN the decade that has elapsed since the discovery of abnormal human haemoglobins a great deal of information about their incidence in various populations has accumulated. Indeed, it is probable that the main facts about the distribution of haemoglobin types distinguishable by the techniques of electrophoresis and chromatography are now known. Recognition of the widespread distribution of the sex-linked gene associated with deficiency of the erythrocyte enzyme glucose-6-phosphate dehydrogenase (the G6PDD gene) has been more recent, but already the broad outlines of distribution of this character are emerging.

The fact that abnormal haemoglobins are confined to certain human populations has inevitably given rise to anthropological speculation, often of a rather unrestrained kind. This meeting of the Society for Human Biology provides a good opportunity for more sober assessment of the anthropological significance of the abnormal haemoglobin and enzyme deficiency traits. This is a matter of general interest in relation to the use of "adaptive" characters in anthropology, for if the abnormal haemoglobin genes —upon which intense selection is known to act—are still useful as population markers, then most other polymorphic characters— upon which selection probably operates less intensely, if at all—are sure to be even more useful.

Polymorphic Haemoglobins

Four abnormal haemoglobin genes attain high frequencies in human populations and fall unambiguously into the definition of polymorphism. The best known is the sickle-cell gene, which attains frequencies of 20 per cent in some East African tribes, is common through most of Central and West Africa, and has also a wide, but irregular, distribution in Sicily, Greece, South Turkey, South Arabia and the Indian peninsula (Fig. 1).

In the common form of thalassaemia heterozygotes have high haemoglobin A_2 levels and the most likely biochemical lesion is retardation of synthesis of the β-polypeptide chain of haemoglobin —which may be abnormal (Ingram and Stretton, 1959). This form is widespread in Mediterranean countries such as Italy, Greece and Turkey, in Middle Eastern countries such as Israel, Iraq and Iran, in India, in South-East Asia and the East Indies (see Chernoff, 1959, for references). There is, unfortunately, little information on the frequency of the thalassaemia gene in most of these areas. The careful studies of Bianco et al. (1952) have shown that thalassaemic heterozygotes attain frequencies of up to 20 per cent in Sardinia and parts of Italy. Olesen et al. (1959) have found that typical thalassaemia heterozygotes, with elevated haemoglobin A_2 levels, are present at a frequency of 11 per cent in Southern Libera. Other isolated reports of this condition in Africans have appeared, but the general impression is that this condition is rather rare in most other African territories. It has not been seen by competent haematologists in East Africa, where another condition interacts with the sickle-cell gene to produce a mild variant of sickle-cell disease. This is a gene which in the heterozygous state is responsible for persistence of foetal haemoglobin into adulthood, without elevation of haemoglobin A_2. The frequency of this gene is not accurately known, but is of the order of 0·1 per cent in Uganda (Jacob and Raper, 1958) and Southern Ghana (Edington and Lehmann, 1955). The same character has been found in subjects of African origin in Jamaica (Went and MacIver, 1958) and the United States (Herman and Conley, 1960).

The third haemoglobin for which human populations are undoubtedly polymorphic is haemoglobin C. Heterozygotes are found in frequencies approaching 30 per cent in some tribes in Northern Ghana (Edington and Lehmann, 1956) and in the adjacent High Volta (Sansarricq et al., 1959). To all sides of this high zone the frequencies of haemoglobin C fall. Only isolated instances of the condition occur to the South-East of the Niger river, although moderate frequencies (up to 2 per cent) occur in some mixed Berbers in Algeria (Cabannes, 1960). The restriction of haemoglobin C to West Africa (apart from very occasional occurrences of apparently identical haemoglobins in such people as the French—see André et al. 1958) raises some interesting problems in population genetics which will be discussed below.

The fourth haemoglobin which is undoubtedly polymorphic is haemoglobin E. As shown in Fig. 3, this haemoglobin has a widespread distribution in South-East Asia. Heterozygote frequencies of up to 35 per cent have been reported in Cambodia (Brumpt et al., 1958), although this may be somewhat over-estimated owing to confusion of high haemoglobin A_2 levels with haemoglobin E in filter-paper electrophoresis. In any case, there is no doubt that both thalassaemia and haemoglobin E are common in many South-East Asian populations (see Luan-Eng, 1958). Occasional cases of haemoglobin E have also been reported from elsewhere, e.g. South Turkey (Aksoy, 1960) and Greece (Gouttas et al., 1960).

Haemoglobins at the Borderline of Polymorphism

Other haemoglobins are found consistently, but at much lower

FIG. 1. Frequency of the sickle-cell gene in the Old World.

%
15-20
10-15
5-10
-5

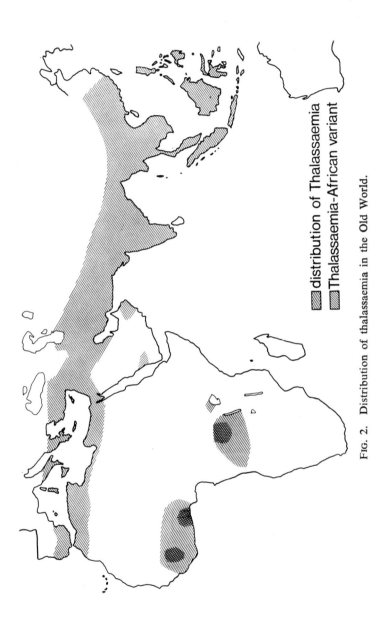

FIG. 2.   Distribution of thalassaemia in the Old World.

■ distribution of Thalassaemia

▨ Thalassaemia-African variant

frequencies, in certain populations. Thus, several Indian populations tested have shown 1–2 per cent of subjects with a haemoglobin having the electrophoretic and chromatographic properties of D, and apparently identical haemoglobins have been described from elsewhere e.g. Algeria (Cabannes, 1960). On detailed investigation, however, the position has turned out to be complicated. Benzer, Ingram and Lehmann (1958) compared three samples of haemoglobin D (from Cyprus, the Punjab and Gujerat). Peptide analysis showed that these haemoglobins are quite distinct from one another.

This illustrates one of the many pitfalls awaiting the anthropologist who tries to use haemoglobins as racial markers. Haemoglobins once thought to be the same are now demonstrably different, and there is reason to believe that the converse also obtains, that haemoglobins once thought to be different are probably identical. Thus, it now seems likely that the abnormal foetal

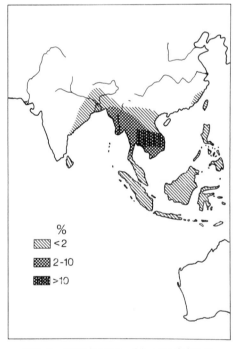

%
▨ < 2
▩ 2-10
▩ >10

Fig. 3.   Distribution of haemoglobin E.

haemoglobin described by Fessas and Papaspyrou (1957) is the same as that subsequently termed Bart's (Ager and Lehmann, 1958). There has also been confusion in terminology of adult haemoglobins having higher anodic mobilities than normal adult haemoglobin at pH 8·6. Haemoglobins now known as K and N are found at frequencies of 0·5 to 2 per cent in many West African tribes (Neel *et al.*, 1956; Sansarricq *et al.*, 1959). Frequencies of up to 13·7 per cent of a haemoglobin apparently identical with K have been described in Algerian Berbers (Cabannes, 1960).

Another example of a haemoglobin at the borderline of polymorphism is the type described by Luan-Eng (1957) as haemoglobin O or Buginese X. This has been reported to be present in about 1 per cent of the Buginese tested (from Sulawesi, what was formerly Celebes), to be absent in other Indonesian populations tested and to be distinct from other known haemoglobins.

Abnormal foetal haemoglobins have already been mentioned. These include Bart's, which is of considerable genetic interest because it consists entirely of γ-polypeptide chains (Hunt and Lehmann, 1959), the Alexandra type (Fessas *et al.*, 1959) and a haemoglobin described as Augusta I which, according to Huisman (1960), consists entirely of sickle-cell β-polypeptide chains. It appears that the abnormal foetal haemoglobins are of only limited interest from the anthropological point of view, since they are rather widely distributed in tropical countries, e.g. Greece, Singapore and Africa (see Hendrickse *et al.*, 1960). Since both Bart's haemoglobin and the adult haemoglobin variant known as H (which consists entirely of β-polypeptide chains) are thought to be associated with the variety of thalassaemia affecting the α-polypeptide chain (Ingram and Stretton, 1959), their widespread distribution in tropical countries in people of different races is not unexpected.

Numerous other haemoglobins recur at still lower frequencies, and are of only limited interest to anthropologists. An account of their distribution is beyond the scope of this paper: Beaven and Gratzer (1959) give a useful summary which was reasonably complete when submitted for publication.

Population Genetics and Distribution of the Sickle-Cell Gene

The distribution of the sickle-cell gene can be understood only in

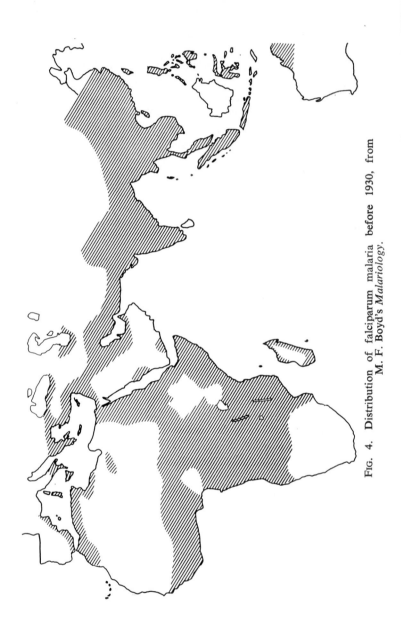

FIG. 4. Distribution of falciparum malaria before 1930, from M. F. Boyd's *Malariology*.

terms of its population genetics. There is now general agreement that in Africa the homozygous condition approaches complete lethality before reproductive age. This may not be the case elsewhere, e.g. in the West Indies (Anderson *et al.*, 1960). Most of the supposed cases of survival of African sickle-cell homozygotes to adulthood turned out on closer analysis to have genetic variants of sickle-cell disease (Edington and Lehmann, 1955; Jacob and Raper, 1958). There is also general agreement that the main factor compensating for loss of sickle-cell genes is the resistance of the heterozygote against malignant tertian malaria. The evidence has recently been reviewed by Allison (1961).

The protection against malaria increases the chances of survival of sickle-cell heterozygotes between birth and reproductive age, and may also increase the fertility of women heterozygous for the sickle-cell gene. The fitness of the sickle-cell homozygote in Africa can be taken as effectively zero. Calculations of the fitness of heterozygotes where the gene is at its maximum frequency of 0·2 show that this is about 25 per cent above that of subjects with normal adult haemoglobin (Allison, 1956). Taking these figures— which are not likely to be greatly in error as representing maximum

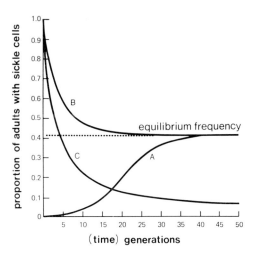

FIG. 5. Rate of change of proportion of subjects carrying sickle-cells in a population under conditions described in the text. A and B, heterozygote favoured. C, no heterozygous advantage.

values—and the equations of Smith (1954), the rates of change of the sickle-cell gene to and from the equilibrium value can be calculated (Fig. 5).

From the point of view of the anthropologist, it is of particular interest to notice that where the heterozygote is at an advantage the initial rise of the sickle-cell frequency is very slow. Thus it would take many centuries in a malarious environment for the heterozygote frequency to rise from 1 per cent to 20 per cent. Moreover, where the heterozygous advantage has disappeared the heterozygote frequencies fall quite rapidly to 10 per cent but more slowly thereafter, approaching the ordinate asymptotically. Thus the gene will persist for many generations as a marker in populations. And this is the main value of the sickle-cell gene from the anthropological point of view. Any attempt to use frequencies of the sickle-cell gene to estimate the contribution of individual populations to mixed populations must be made with extreme caution.

An instructive comparison of the frequencies of the sickle-cell and haemoglobin C genes in Dutch West Indian territories has been published by Jonxis (1959) and van der Sar (1959). In the seventeenth century the Dutch maintained a "factory" at Elmina on the Gold Coast. Most of the slaves transported to their New World possessions were probably derived from there. Whether the slave ships went to Surinam or Curaçao appears to have been a matter of chance, so that the slave population in the two territories probably had similar origins. Surinam has been intensely malarious, but not Curaçao. Three groups of Negroes studied in the former showed relatively high frequencies of the sickle-cell trait (about three times that of the haemoglobin C trait). In Curaçao, on the other hand, sickle-cell trait frequencies are low—less than those of haemoglobin C. The difference in frequency of the sickle-cell gene in the two localities is statistically highly significant. It cannot be explained by Caucasian or Indian admixture, which from blood-group studies appears to be rather slight in Curaçao and is probably even less in Surinam.

Jonxis plausibly interprets the finding as evidence that the advantage of the sickle-cell heterozygote remained in Surinam but not in Curaçao. But if the origins and environmental factors affecting these two groups had not been known, it might well have been concluded that they came from different parts of Africa. The fact that powerful selective effects operate on the sickle-cell gene makes nonsense of any attempt to use the character for large-scale tracing of racial origins and movements. At a time when the sickle-cell gene had been shown to occur among the pre-Dravidians in the Nilgiri hills of South India, and to attain higher frequencies in East than in West Africa, Lehmann and Cutbush (1952) took the evidence as support for the idea of an Indian migration to Africa in prehistoric times. This hypothesis was later restated (Lehmann, 1954) to give a common origin for the sickle-cell genes of Asia, Africa and Europe. He suggested that sickle-cell mutants were first established in the population of the Arabian peninsula, whence they were spread by Veddoid emigrants to India, Africa and Europe. This diffusionist hypothesis has been elaborated by Brain (1953) and Singer (1954), who suggested that the postulated immigrants to Africa were the cattle-herders who imported Zebu stock from Asia at the beginning of the present era.

All this may or may not be true, but the incidence of the sickle-cell trait in these populations has very little bearing on the problem, as later work has shown. The trait is by no means confined to the pre-Dravidians of South India. It is also common among certain "scheduled castes" elsewhere in India and in labourers brought into Assam from Orissa (Dunlop and Mozumber, 1952; Bhatia *et al.,* 1955; Shukla *et al.,* 1958; Chatterjea, 1959). The Greek and Turkish populations having high frequencies of the sickle-cell trait do not have more of the characteristically African blood groups than their compatriots, so there is nothing to suggest that they have an unusual degree of African admixture. There are other explanations for the lower frequencies of the sickle-cell gene in West Africa than late diffusion from East Africa, and these will be discussed more fully below.

It is generally accepted that the sickle-cell gene is produced by an uncommon mutation of the gene controlling synthesis of the β-polypeptide chain of haemoglobin. The upper limit for the mutation rate suggested by the observations of Vandepitte *et al.* (1954) is for various reasons certainly too high, and we can do no more than speculate about the actual mutation rate at this locus. If it is 1×10^{-7}—a mean figure based on mutation rates in other species and on what little can be estimated of mutation rates in man—

then some 10 sickle-cell genes will arise in malarious parts of Africa per generation. We do not know, and indeed cannot ascertain, whether the sickle-cell gene arose independently by mutation in the African, European and Asian populations in which it is now found, or was transferred by movement of relatively small numbers of subjects from one group to another. What we can be sure about is that in the absence of selective advantage the gene could not have attained the observed frequencies. It is no accident that high frequencies are only found in regions where malaria is— or was until recently—endemic. Furthermore, the presence of the sickle-cell gene in a mixed population has limited value as a guide to population origins, since it could imply contribution from one or more of many population groups (African, Greek, Turkish, Arab or Indian).

Within regions the situation is not much clearer. Lehmann and Raper (1949) originally argued that high frequencies of the sickle-cell trait in Uganda indicated the presence of Nilotic blood whereas low frequencies were indicators of Hamitic blood; in Bantu tribes the frequency of the sickle-cell trait was supposed to be inversely related to the degree of Hamitic admixture. When the importance of malaria in maintaining the frequency of the sickle-cell gene was recognized, an alternative explanation could be offered (Allison, 1954): that on the whole the Hamites lived in non-malarious regions, the Nilotes in malarious regions and the Bantu in both. Where people of Hamitic origin (e.g. the Teso) live in malarious country, they are found to have high frequencies of the sickle-cell trait. In the whole of East Africa there was, in fact, an excellent correspondence between the distribution of the sickle-cell trait and malaria: trait frequencies of above 10 per cent being found only where falciparum malaria is hyperendemic or holoendemic. This conclusion still stands, with some minor qualifications. Thus, although the intensity of malarial transmission is about the same in the adjacent Kambe and Duruma tribes of Kenya, the sickling rate in samples of the former were found to be 33 per cent and in the latter only 9 per cent (Foy *et al.*, 1954). Even more striking variations have been found by Dr D. F. Clyde and myself within a single small tribe, the Bondei, in Tanganyika. Here the sickle-cell trait frequency varies from 7 per cent to 40 per cent in different villages. In the Southern Sudan falciparum malaria is hyperendemic but the sickling rate among the Northern Nilotes is reported to be generally less than 5 per cent (Roberts and Lehmann, 1955).

In West Africa the distribution of the sickle-cell gene is complicated by a number of factors. In general, as in East Africa, high frequencies are found only where malaria is hyperendemic or holoendemic, although there is one exception: sickle-cell trait frequencies of over 30 per cent have been reported from an epidemic area near Lake Chad (Roberts, Lehmann and Boyo, 1960). The first complication can now be considered. This is the presence of haemoglobin C in high frequencies in parts of West Africa. Because individuals who inherit both the sickle-cell and haemoglobin C genes are at a disadvantage, Allison (1955) pointed out that the two genes should tend to be mutually exclusive in human populations. This was subsequently found to be the case in Ghana and Sierre Leone (Allison, 1956; Edington and Lehmann, 1956) and Nigeria (Garlick, 1960) where there are very high frequencies of the sickle-cell trait, and the sum of the two genes in populations is nearly the same over the whole region. Hence the three genes (A, S and C) could be in or near genetic equilibrium.

This is clearly not the case in Liberia, where there are populations in highly malarious environments having low frequencies of both S and C haemoglobins (Neel et al., 1956). Livingstone (1958) has argued that the sickle-cell gene has only recently reached the forest lands in and near Liberia, and that it is still spreading from North to South and East to West in that region. He argues that the present high density of population in the forest zone of West Africa dates from the recent introduction of iron tools for forest clearance and of yam cultivation for the efficient exploitation of forest soils. He suggests, further, that the population density would have been too low and potential breeding places for anophelines in the forest too few to support holoendemic malaria. The populations with low sickling frequencies in Portuguese Guinea and Liberia would then be the remnants of the pre-neolithic or "palaeonegroid" population.

The hypothesis that sickling spread into the West African forests comparatively recently is plausible, but the evidence linking this spread with the introduction of iron tools and forest agriculture is by no means conclusive. As Garlick (1960) has recently emphasized, stone tools and burning have been used efficiently for forest clearance elsewhere, and the postulated increase in productivity following introduction of iron tools may well have been exaggerated. How and when the yam and cocoyam became available to cultivators in different parts of West Africa are quite obscure. Furthermore, the minimum size of community necessary to support malaria is probably small—as the presence of malaria in anthropoids and pygmies shows—and *Anopheles funestis* serves as an efficient vector of malaria in West Africa where it is too heavily wooded for *A. gambiae* to breed. In the French territories near Liberia there is no consistent relationship between low S and C frequencies and people classified by anthropologists as palaeonegroid (Sansarricq *et al.*, 1959). And the finding of relatively high frequencies of thalassaemia in Southern Liberia (Olesen *et al.*, 1959) may well be relevant to the low frequencies of the sickle-cell gene there. Individuals who are heterozygous for the thalassaemia and sickle-cell genes have a disease that is usually more severe than haemoglobin C: sickle-cell disease; and it follows that if the thalassaemia gene is present in a population the sickle-cell gene will spread only slowly, if at all, in that population.

In summary, it is clear that the distribution of the sickle-cell gene in West Africa cannot be explained in any simple way. The sickle-cell and haemoglobin C genes could be in or near a stable equilibrium in Nigeria, Ghana and the adjacent part of High Volta, but it is by no means certain that this is so. In Liberia and Portuguese Guinea other factors, including the presence of thalassaemia in fairly high frequencies and probably population migrations, contribute to the low S and C haemoglobin frequencies, but the relative importance of these and other factors is still a matter for speculation. On Livingstone's attempt to link the introduction of the sickle-cell gene into the West African forests with the immigration of particular Iron Age populations a verdict of "not proven" must be returned.

Distribution of Other Abnormal Haemoglobins

If the distribution of the sickle-cell gene in West Africa is still rather confusing, that of the other abnormal haemoglobins is even more confusing. Why the haemoglobin C gene can attain such high frequencies in Northern Ghana and the nearby High Volta is unknown. The limited available information suggests that the haemoglobin C homozygote is at a slight disadvantage, but that subjects with haemoglobin C: sickle-cell disease are at a greater disadvantage (having crises in pregnancy and other disabilities). From considerations of population genetics (Allison, 1956) the only plausible conclusion is that haemoglobin C heterozygote is favoured in West Africa. The only investigation so far carried out on rather small numbers (Edington and Laing, 1957) has not shown this genotype to have resistance to falciparum malaria. For equilibrium a much smaller advantage than that of the sickle-cell heterozygote would suffice, and it might be difficult to obtain unequivocal evidence of its existence in field studies. The distribution of the trait in West Africa suggests that protection against other species of malaria parasites may occur (Livingstone, 1960), and further investigations of this problem would be worth while. The fall in frequency of haemoglobin C, but not S, in Surinam (Jonxis, 1959) also suggests that different environmental agencies may affect the two genes.

In an attempt to explain the restricted distribution of haemoglobin C in West Africa, Allison (1956) mentioned two possibilities. The first is that some environmental factor present in West Africa, but not East Africa, favours the haemoglobin C heterozygote. The second is that by chance the haemoglobin C gene was established in or near Northern Ghana before the sickle-cell gene, thereby excluding the latter. We are still in no position to decide between these alternatives. Livingstone (1958) has argued that high haemoglobin C frequencies coincide with the distribution of the Gur sub-family of the Niger-Congo language group. While this appears to be true of the highest frequencies, moderately high frequencies of haemoglobin C are widespread among Kwa-speakers to the South.

The thalassaemia gene is just as lethal as the sickle-cell gene when homozygous, and yet heterozygote frequencies up to 20 per cent have persisted in parts of Italy, Greece and Asia (Fig. 2). The heterozygote must therefore be (or have been until recently) at a considerable advantage. Once again the local distribution in countries such as Sardinia (Ceppellini, 1955), as well as the overall distribution in world populations, strongly suggests that malaria is involved. As outlined above, in several South-East Asian countries, such as Burma, Thailand and Indonesia, both the thalassaemia and haemoglobin E genes are common. The latter is probably disadvantageous only rarely when homozygous, but there is no denying the disadvantage of individuals who inherit both the haemoglobin E and thalassaemia genes. The only plausible inference is that the possession of a single abnormal gene is advantageous, and the distribution of haemoglobin E again brings malaria to mind.

The question then arises why the sickle-cell gene is not found in South-East Asia. The mutual exclusion principle discussed above seems the most likely explanation: that mutations to thalassaemia and haemoglobins S and E recur at low frequencies in all populations, but that where malaria is common one or two of these genes became established. It would then be difficult for other genes to spread.

Cattle Haemoglobins

Cattle also have electrophoretically distinct haemoglobin types. A haemoglobin type known as bovine B has been described in Algerian, French and Channel Island cattle breeds. This haemoglobin is also found in Nigerian and East African zebu cattle, and at lower frequencies in the Hamitic longhorn Ankole cattle (Bangham and Blumberg, 1958; Lehmann and Rollinson, 1958). The relatively high incidence of Bovine B haemoglobin in Indian Gir cattle has been taken by Lehmann (1959) as an indication of a substantial genetic contribution from *Bos indicus,* and as evidence of a close relationship between Indian and African zebu cattle. While such a relationship is likely on purely morphological grounds, the frequency of bovine B haemoglobin may well be unreliable as an indicator of relationship. Indeed in the sample of Gir cattle tested the proportion of heterozygotes was nearly double that expected from the random mating hypothesis. If this remarkable observation is confirmed on other samples it will imply very considerable heterozygous advantage—much greater than that of the sickle-cell heterozygote. Who is to say, then, that the distribution of cattle haemoglobins today is not determined primarily by environmental selective agencies? To use the presence of bovine B haemoglobin frequencies as an indicator of migration of cattle, and associated human populations, from the Indian sub-continent seems premature in the present state of knowledge.

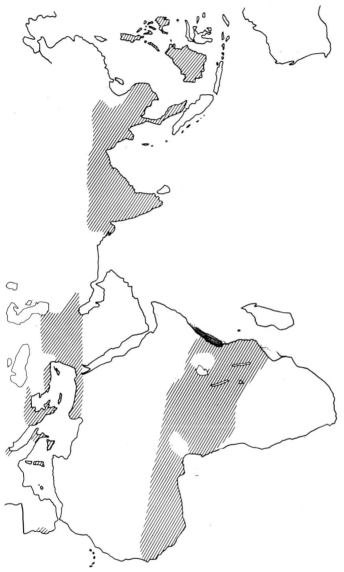

FIG. 6. Known distribution of glucose-6-phosphate dehydrogenase deficiency in frequencies above 2 per cent in Old World males.

Glucose-6-Phosphate Dehydrogenase Deficiency (G6PDD)

It has been shown (Beutler, 1959) that about 11 per cent of American Negro males have an intrinsic red blood cell abnormality. A striking feature of the abnormality is low activity of the enzyme glucose-6-phosphate dehydrogenase, which is responsible for the first and rate-controlling step in the hexose-monophosphate-shunt metabolic pathway. The primary genetically-controlled abnormality may not be defective synthesis of the enzyme itself but absence of a stromal factor that activates the enzyme (Rimon et al., 1960).

The defect is inherited as a sex-linked character with full expression in hemizygous males and partial expression in heterozygous females. Persons with the defect are liable to develop haemolysis neonatally, on exposure to therapeutic doses of the antimalarial 8-aminoquinolines (primaquine, pamaquine, etc.) and certain other drugs, on ingestion of broad beans (*Vicia faba*) and when they have certain virus diseases (Marks, 1960).

It is now becoming apparent that the G6PDD trait has a widespread distribution in tropical countries (Fig. 6). Motulsky (1960) and Allison (1960) have pointed out that the distribution of the trait corresponds with that of malignant tertian malaria. The local distribution also favours this view: in East Africa and Sardinia some areas are, or were until recently, very malarious and others free from the disease. High frequencies of the enzyme-deficiency trait are found only in malarious regions. On theoretical grounds it might also be expected that the enzyme-deficiency trait could limit multiplication of malaria parasites (see Allison, 1961). A direct field test of the malaria protection hypothesis was carried out in collaboration with Dr. D. F. Clyde in Tanganyika. Male children with enzyme deficiency between four months and four years of age had significantly lower parasite counts than children with normal enzyme activities: enzyme-deficient female children (mostly heterozygotes) had fewer parasites than those with normal enzyme activities, but the results were not statistically significant.

We can therefore make some speculations about the population genetics of the G6PDD trait, which is one of the most striking examples of a sex-linked polymorphism in any animal form. The heterozygous female is at an advantage in malarious regions while the male hemizygote—being liable to haemolysis neonatally, from virus diseases and perhaps other causes—may actually be at a net disadvantage. This is the simplest way in which a stable equilibrium at a sex-linked locus can be maintained (Mandel, 1959). If the enzyme-deficient hemizygote and homozygote were at a net advantage the deficiency gene might have replaced the normal allele in some population, whereas the frequency of the former rises to 0·3 in many regions but is not known to exceed 0·4.

Because of its widespread distribution in the tropics, the G6PDD gene is of limited value as a racial marker. Its value in this respect is somewhat increased by the fact that there seem to be two distinct varieties of the mutant (Marks and Gross, 1959). Caucasians with the defect have significantly lower mean enzyme levels than those of deficient Negroes, and the latter show greater differences between enzyme activities of young and old cells.

Conclusion

From what has been said, my own view of the distribution of the abnormal haemoglobins and G6PDD traits will be clear: that these are determined largely by selection resulting from environmental agencies, in particular malaria, and only secondarily by population movements. Because of what might be termed genetic inertia—the relatively slow rate of change of gene frequencies in populations—these characters have some value as racial markers. Thus, if one found a population with significant frequencies of the S, C and G6PDD genes (such as West Indian and American Negroes) it would be reasonable to infer that they were related to the West African Negroes. If one found a white population with significant frequencies of haemoglobin K one would look for a link with Algerian Berbers. The presence of thalassaemia, haemoglobin E and G6PDD would suggest strongly South-East Asian links. But these conclusions are hardly very startling. I am not aware of a single instance where the distribution of the abnormal haemoglobins has pointed the way to some new or unexpected relationship. And the selection of certain facts about abnormal haemoglobin distribution to support particular interpretations of population movements belongs more properly to the realm of after-dinner speculation than to that of objective science.

The great value of abnormal haemoglobin studies has been that they have changed the whole climate of opinion about genetically-determined characters in man. We now expect polymorphic characters to have selective values, and are often disappointed when there is no indication as to the nature of the selective agencies operating on them. This is the justification for the intensive effort that is currently being expended on determination of the frequency of polymorpric characters in human populations. When large numbers of facts are assembled, it may be possible to draw more general conclusions—if only because selection is operating more or less independently on the different systems.

References

AGER, J. A. M. and LEHMANN, H. (1958) Observations on some "fast" haemoglobins: K, J. N and "Bart's". *Brit. Med. J.* **2**, 142.

AKSOY, M., (1960) The haemoglobin E syndromes. Haemoglobin E in Eti-Turks. *Blood* **15**, 606.

ALLISON, A. C. (1954) The distribution of the sickle-cell trait in East Africa and elsewhere, and its apparent relationship to the incidence of subtertian malaria. *Trans. Roy. Soc. Trop. Med. Hyg.* **48**, 312.

ALLISON, A. C. (1955) Aspects of polymorphism in man. *Cold Spr. Harb. Symp. Quant. Biol.* **20**, 239.

ALLISON, A. C. (1956) The sickle-cell and haemoglobin C genes in some African populations. *Ann Hum. Genet.* **21**, 67.

ALLISON, A. C. (1960) Glucose-6-phosphate dehydrogenase deficiency in red blood cells of East Africans. *Nature (Lond.)* **186**, 431.

ALLISON, A. C. (1961) Genetic factors in resistance to malaria. *Trans. N.Y. Acad. Sci.* **91**, 710.

ANDERSON, M., WENT, L. N., MACIVER, J. E. and DIXON, H. G. (1960) Sickle-cell disease in pregnancy. *Lancet* **ii**, 517.

ANDRÉ, R., BESSIS, M., DREYFUS, B., JACOB, S. and MALASSENET, R. (1958) Association d'un syndrome thalassémique et d'une hémoglobine anormale à l'electrophorèse non encore identifiée, chez un Français d'origine picarde. *Rev. d'Hémat.* **13**, 31.

BANGHAM, A. D. and BLUMBERG, B. S. (1959) Distribution of electrophoretically different haemoglobins among some cattle breeds of Europe and Africa. *Nature (Lond.)* **181**, 1551.

BEAVEN, G. H. and GRATZER, W. B. (1959) A critical review of human haemoglobin variants. *J. Clin. Path.* **12**, 101.

BEUTLER, E. (1959) The hemolytic effect of primaquine and related compounds: a review. *Blood* **14**, 103.

BENZER, S., INGRAM, V. M. and LEHMANN, H. (1958) Three varieties of haemoglobin D. *Nature (Lond.)* **182**, 852.

BHATIA, H. M., THIN, J., DEBRAY, H. and CABANNES, J. (1955) Étude anthropologique et génétique de la population du nord de l'Inde. *Bull. Soc. Anthrop. Paris.* **6**, 199.

BIANCO, I., MONTALENTI, G., SILVESTRONI, E. and SINISCALCO, M. (1952) Further data on the genetics of microcythaemia or thalassaemia minor and Cooley's disease or thalassaemia major. *Ann. Eugen. (Lond.)* **16**, 299.

BRAIN, P. (1953) The sickle-cell trait: a possible mode of introduction into Africa. *Man.* **53**, 154.

BRUMPT, L. and V., COQUELET, M. and DE TRAVERSE, P. M. (1958) La detection de l'hémoglobine E. Étude des populations cambodgiennes. *Rev. d'hémat.* **13**, 21.

CABANNES, R. (1960) The distribution of abnormal haemoglobins in Algeria, the Hoggar and High Volta: anthropological incidences. *J. Roy. Anthrop. Inst.* **90**, 307.

CEPPELLINI, R. (1955) Discussion of "aspects of polymorphism in man". *Cold Spr. Harb. Symp. Quant. Biol.* **20**, 239.

CHATTERJEA, J. B. (1959) Haemoglobinopathy in India. In *Abnormal Haemoglobins.* Blackwell, Oxford.

CHERNOFF, A. I. (1959) The distribution of the thalassaemia gene: a historical review. *Blood* **14**, 899.

DUNLOP, K. J. and MOZUMBER, U. K. (1952) Occurrence of sickle-cell anaemia among a group of tea garden labourers in Upper Assam. *Ind. Med. Gaz.* **87**, 387.

EDINGTON, G. M. and LAING, W. N. (1957) Relationship between haemoglobins C and S and malaria in Ghana. *Brit. Med. J.* **2**, 143.

EDINGTON, G. M. and LEHMANN, H. (1955) Expression of the sickle-cell gene in Africa. *Brit. Med. J.* **2**, 1328.

EDINGTON, G. M. and LEHMANN, H. (1956) The distribution of haemoglobin C in West Africa. *Man* **56**, 34.

FESSAS, PH. and PAPASPYROU, A. (1957) New "fast" hemoglobin associated with thalassemia. *Science* **126**, 1119.

FESSAS, PH., MASTROKALOS, N. and FOSTIROPOULOS, G. (1959) New variant of human foetal haemoglobin. *Nature (Lond.)* **183**, 30.

FOY, H., KONDI, A., TIMMS, G., BRASS, W. and BUSHRA, F. (1954) The variability of sickle-cell rates in the tribes of Kenya and the Southern Sudan. *Brit. Med. J.* **1**, 294.

GARLICK, J. P. (1960) Blood groups and sickling in Nigeria. Ph.D. Thesis: University of London.

GOUTTAS, A., TSEVRENIS, C., ROMBOS, C., PAPASPYROU, A. and GARIDI, M. (1960) L'hémoglobinose E en Grèce. *Le Sang.* **31**, 1.

HENDRICKSE, R. G. BOYO, A. E. FITZGERALD, P. A. and KUTI, S. R. (1960) Studies on the haemoglobin of newborn Nigerians. *Brit. Med. J.* **i**, 611.

HERMAN, E. C. and CONLEY, C. L. (1960) Hereditary persistence of foetal haemoglobin: a family study. *Amer. J. Med.* **29**, 9.

HUNT, J. A. and LEHMANN, H. (1959) Haemoglobin "Barts", a foetal haemoglobin without α-chains. *Nature (Lond.)* **184**, 872.

HUISMAN, T. H. J. (1960) Properties and inheritance of the new fast haemoglobin type found in umbilical cord blood samples of Negro babies. *Clin. Chim. Acta* **5**, 709.

INGRAM, V. M. and STRETTON, A. O. W. (1959) Genetic basis of thalassaemia diseases. *Nature (Lond.)* **184**, 1903.

JACOB, G. F. and RAPER, A. B. (1958) Hereditary persistence of foetal haemoglobin production and its interaction with the sickle-cell trait. *Brit. J. Haematol.* **4**, 138.

JONXIS, J. W. P. (1959) The frequency of haemoglobin S and haemoglobin C carriers in Curaçao and Surinam. In *Abnormal Haemoglobins.* Blackwell, Oxford.

LEHMANN, H. (1954) Distribution of the sickle-cell gene. A new light on the origin of East Africans. *Eugen. Rev.* **46**, 101.

LEHMANN, H. (1959) The haemoglobins of 103 Indian cattle. *Man* **59**, 62.

LEHMANN, H. and CUTBUSH, M. (1952) Sickle-cell trait in Southern India. *Brit. Med. J.* **i**, 104.

LEHMANN, H. and ROLLINSON, D. H. L. (1958) The haemoglobins of 211 cattle in Uganda. *Man* **57**, 52.

LEHMANN, H. and RAPER, A. B. (1949) The distribution of the sickle-cell trait in Uganda, and its ethnological significance. *Nature (Lond.)* **164**, 494.

LIVINGSTONE, F. B. (1958) Anthropological implications of sickle-cell gene distribution in West Africa. *Amer. Anthrop.* **60,** 533.
LIVINGSTONE, F. B. (1960) Private communication.
LUAN-ENG, L. I. (1957) A new haemoglobin in the Buginese. *Lancet* **ii,** 1338.
LUAN-ENG, L. I. (1959) Pathological haemoglobins in Indonesia. In *Abnormal Haemoglobins.* Blackwell, Oxford.
MANDEL, S. P. H. (1959) Stable equilibrium at a sex-linked locus. *Nature (Lond.)* **183,** 1347.
MARKS, P. A. (1960) *in Proc. 1st Macey Conference on Genetics,* ed. SUTTON, H. E., 200–213.
MARKS, P. A. and GROSS, R. T. (1959) Erythrocyte glucose-6-phosphate dehydrogenase deficiency : evidence of differences between Negroes and Caucasians with respect to this genetically-determined trait. *J. Clin. Invest.* **38,** 2253.
MOTULSKY, A. G. (1960) Metabolic polymorphisms and the role of infectious diseases in human evolution. *Hum. Biol.* **32,** 28.
NEEL, J. V., HIERNAUX, J., LINHARD, J., ROBINSON, A., ZUELZER, W. W. and LIVINGSTONE, F. B. (1956) Data on the occurrence of haemoglobin C and other abnormal haemoglobins in some African populations. *Amer. J. Hum. Genet.* **8,** 138.
OLESEN, E. B., OLESEN, K., LIVINGSTONE, F. B., COHEN, F., ZUELZER, W. W., ROBINSON, A. R. and NEEL, J. V. (1959) Thalassaemia in Liberia. *Brit. Med. J.* **1,** 1385.
RIMON, A., ASHKENASI, I., RAMOT, B. and SHEBA, C. (1960) Activation of glucose-6-phosphate dehydrogenase of enzyme-deficient subjects. 1. Activation of stroma of normal erythrocytes. *Biochem. and Biophys. Res. Comm.* **2,** 138.
ROBERTS, D. F. and LEHMANN, H. (1955) A search for abnormal haemoglobins in some southern Sudanese peoples. *Brit. Med. J.* **i,** 519.
ROBERTS, D. F., LEHMANN, H. and BOYO, A. E. (1960) Abnormal haemoglobins in Bornu. *Amer. J. Phys. Anthop.* n.s. **18,** 5.
SANSARRICQ, H., MARILL, G., PORTIER, A. and CABANNES, R. (1959) Les hémoglobinopathies en Haute-Volta. *Le Sang* **30,** 503.
SHUKLA, R. M., SOLANKI, B. R. and PARANDE, A. S. (1958) Sickle-cell disease in India.
SINGER, R. (1954) Origin of the sickle-cell gene. *S. Afr. J. Sci.* **50,** 287.
SMITH, S. M. (1954) Appendix to notes on sickle-cell polymorphism. *Ann. hum. Genet.* **21,** 83.
VANDEPITTE, J. M., ZUELZER, W. W., NEEL, J. V. and COLAERT, J. (1955) Evidence concerning the inadequacy of mutation as an explanation of the frequency of the sickle-cell gene in the Belgian Congo. *Blood* **10,** 341.
VAN DER SAR, A. (1959) The occurrence of carriers of abnormal haemoglobin S and C on Curaçao. Groningen : Dykstra's Drukkery.
WENT, L. H. and MACIVER, J. E. (1958) An unusual type of hemoglobinopathy resembling sickle-cell-thalassaemia disease in a Jamaican family. *Blood* **13,** 559.

HAPTOGLOBINS AND TRANSFERRINS

N. A. BARNICOT

For the serum proteins, there is a similar situation to that for the red cell enzymes discussed in the previous paper. For the haptoglobins (e.g. Sutton 1970) the chemical structure of the genetic variants has been identified, the major division of Hpl into fast and slow subtypes has been made. There has been good progress with transferrin structure. But a variety of other serum proteins (such as the group specific component, the components of complement) are now known to have variant genetic forms, and the patterns of world distribution have begun to emerge to varying extents. There also has appeared the curious interaction of the Hp phenotypes with the ABO blood groups giving rise to some segregation distortion, the interpretation of which is not yet clear.

IT IS hardly surprising that haematology has played such an important part in the more recent developments in human genetics. Blood is an easy tissue to sample and blood specimens for laboratory investigation are taken on an enormous scale in medical work. The erythrocytes are highly specialized, but also relatively simple, free-floating cells which are thus most convenient in testing for surface antigens and they contain mainly haemoglobin which is easily liberated for chemical analysis, together with various enzymes either free or attached to the stroma. The plasma on the contrary is a highly complex fluid containing an enormous variety of dissolved substances, including a wide range of proteins in very varying concentrations and with many different functions. The current view that gene action is fundamentally concerned with protein synthesis suggests that a great wealth of genetically controlled variability may await discovery in the plasma or serum, and modern techniques for the separation and characterization of proteins have now made the detection of small individual variations in specific components feasible. I shall deal with two classes of serum proteins in which simply inherited variations, producing no obvious clinical defect, and of common occurrence in many populations, have lately been discovered.

Haptoglobins

Rather more than twenty years ago when Jayle and his colleagues in Paris were studying the peroxidase activity of haemoglobin solutions they noticed that it was enhanced by the addition of serum. They found that this was due to the haemoglobin combining irreversibly with certain proteins of the α_2-globulin fraction which were therefore named haptoglobins. In normal human serum the α_2 fraction comprises some 10 per cent of the total protein and about one-quarter of this fraction is haptoglobin. Haptoglobins are glycoproteins with a relatively high carbohydrate content. Their high affinity for haemoglobin ensures that free haemoglobin does not occur in the plasma until these carriers are saturated. The bond with haemoglobin depends on the globin rather than the haeme but is not very highly specific since human haptoglobin can combine not only with HbA, but also with HbF, variants such as HbS and HbC and even some animal haemoglobins.

In 1955 Smithies in Toronto described a method for electrophoretic separation of proteins in a starch-gel. The resolution of human serum by this new method is superior to that in open-boundary or paper electrophoresis and additional components can be seen. Smithies noticed that the pattern of protein bands was in certain respects different in different individuals, and by adding haemoglobin to the sera he showed that the variable bands were haptoglobins.

The haptoglobin patterns are seen most clearly when haeme-containing regions of the gel are selectively stained with benzidine or o-dianisidine reagents. Smithies recognized three types of pattern (Fig. 1): the first, now known as type 1-1, has a single strong haptoglobin band running well forward near the main β-globulin band. In type 2-2, on the other hand, this band is absent but there are a number of weaker ones nearer the starting-point. In the third type, 2-1, there is a rather weak band at the same position as that in type 1-1, and also a number of slower bands which do not correspond to those in type 2-2 either in position or in their relative strengths.

As a result of family studies Smithies and Walker (1956) suggested that these three types were determined by a pair of allelic genes Hp^1 and Hp^2, and later work (Galatius-Jensen, 1958a; Harris *et al.*, 1959b; Giblett and Steinberg, 1960) on much more extensive material has confirmed this, though a few curious examples of abnormal segregation have been reported (Harris *et al.*, 1958b). At least four additional phenotypes have also been described.

There have been some technical innovations improving the

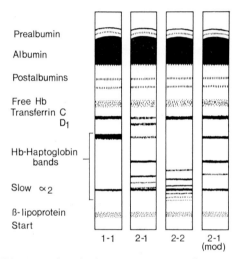

FIG. 1. Diagram of main human serum protein components as seen in a horizontal starch-gel run using Poulik's discontinuous buffer system. Excess HbA has been added to the sera. The diagram does not show the γ-globulin much of which runs behind the starting-point.
Four haptoglobin phenotypes (1-1, 2-1, 2-2, 2-1M) are represented and also a slow transferrin variant (TfD$_1$).

starch-gel method and enlarging its scope. Irregularities due to the method of inserting sera in the gel on filter-paper or mixed with starch grains were overcome (Smithies, 1959a) by introducing fluid serum into a slot and running the gel in the vertical position. By this method a series of as many as ten bands of progressively decreasing strength can be seen in type 2-2 sera. If the gel is made in tris-buffer and borate buffer is used in the tanks (Poulik, 1957) a junction line of steep potential gradient migrates through the gel and the first Hp band is well separated from any excess of free haemoglobin and from the iron-binding β-globulin or transferrin which may also show variants of differing mobility. A preliminary separation of the serum into albumin, α_1, α_2, β and γ-globulins by paper-electrophoresis can be followed by starch-gel electrophoresis of the paper strip. By this two-dimensional method some thirty components have been detected (Smithies, 1959b) (Fig. 2).

The proteins separated in starch-gel can also be examined by so-called immunoelectrophoresis. They are allowed to diffuse into agar where they meet antiserum diffusing in the opposite direction and form precipitation bands at points where reaction occurs. Specific antisera can thus be used to identify particular proteins and antisera to whole human serum can be employed in a search for new components (Hirschfeld, 1959, 1960).

It seems distinctly odd that type 2-2, the homozygote for the gene Hp², should show not one but many protein bands and it prompts the question how the gene controls the production of more

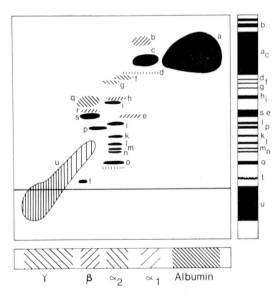

Fig. 2. Two dimensional separation (From Smithies, O., in *Recent Advances in Protein Chemistry,* Vol. 14, p. 65, 1959. Fig. 5, p. 83. By permission of The Academic Press.)

than a single molecule species. A further problem is posed by type 2-1, the heterozygote, since the bands are not only multiple but different from those in type 2-2. The multiple components are not artefacts since when they are isolated and then rerun they retain their characteristics, and the haptoglobin patterns of whole serum persist over a wide range of buffer conditions (Smithies and Connell, 1959).

By paper-electrophoresis the three types can be distinguished by slightly differing mobilities but each moves as an essentially single band (Fig. 3). The reason why multiple components show up in starch-gel appears to be that the rate of migration in this medium is greatly influenced by the size of the molecule. It appears, then, that types 2-1 and 2-2 contain a series of proteins of varying molecular size, and this view is consistent with evidence from ultracentrifuge experiments (Bearn and Franklin, 1959).

A theory which simplifies our conception of the essential action of the haptoglobin genes has been formally set out by Allison (1959b). On this view the gene Hp^2 produces a single protein which forms a series of stable polymers; the protein produced by Hp^1 does not polymerize in this way when present alone, but will combine with the Hp^2 protein to form a different polymer series. The production of multiple proteins is thus seen as secondary to the synthesis of two primary species. The isolation of these and their combination *in vitro* to form the type 2-1 series has not so far been reported, but Smithies and Connell (1959) have given a preliminary account of work in which various haptoglobin phenotypes were split with urea and thioglycollic acid. Type 1-1 yielded a single component of changed mobility while other types not only yielded an apparently identical component to this but also fast-moving ones which differed in each case. These experiments seem to indicate that various haptoglobin proteins have a major constituent in common.

I have already mentioned the discovery of other haptoglobin phenotypes. One of these (Connell and Smithies, 1969) which is fairly common in Africans and American Negroes (Giblett, 1959) is called type 2-1 (mod), or 2-1M, because it looks like a modification of the usual 2-1. The first Hp band is unusually strong and is followed by a rather weaker second and an even fainter third, both of which correspond in position to those in type 2-1 (Fig. 1).

Pedigrees from Europe, where this type is uncommon, suggest that it is inherited, but the most extensive data are those of Giblett and Steinberg (1960) from American Negro families. Their evidence supports the view previously advanced by others (Galatius-Jensen, 1958b), that 2-1M may be due to a variant of the Hp² gene, which may be written Hp²ᵐ. Thus matings of 1-1 × 2-1M give 2-1M children but not 2-1 and so do 2-1M × 2-1 Mmatings.

Several workers (Allison *et al.*, 1958; Harris *et al.*, 1959a, b; Barnicot *et al.*, 1960) have found that in sera from some African populations 30 per cent or more may show no benzidine positive bands. This condition is unusual in normal Europeans but it occurs

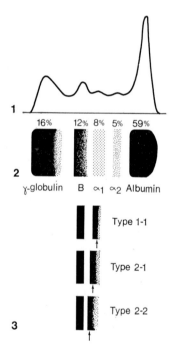

FIG. 3.   (1) Separation of human serum proteins by Tiselius electrophoresis. Veronal pH 8·6.
(2) Separation of same serum by paper electrophoresis.
(3) Positions of free HbA and of haptoglobin of three different phenotypes in paper electrophoresis.
(1 and 2 based on Gerald R. Cooper, Fig. I, p. 57 in *The Plasma Proteins,* ed. Putnam, Vol. I, 1960. By permission of The Academic Press.)

in low but appreciable frequency in American Negroes (Giblett, 1959) and in various other populations. The scoring obviously depends on the sensitivity of the method used and specimens with weak and very weak reactions no doubt grade into haptoglobin-negatives. In Europe, where this phenotype is rare, the occurrence of several cases in a pedigree suggests a genetical basis. It is known (Nyman, 1959), however, that plasma haptoglobin is often greatly depleted or undetectable in various haemolytic diseases; experimental injection of haemoglobin leads to its rapid disappearance followed by a return to normal levels in about a week. Increased red-cell destruction due to malaria or other causes may, therefore, be an important cause of Hp depletion in some tropical peoples. Studies on Northern Nigerian families (Barnicot *et al.*, 1960) showed that the frequency of haptoglobin-negatives was not significantly different in various mating types, which suggested that, at least in this population, environmental rather than genetic causes predominate. The incidence of malarial infections in young children from the area was high, exceeding that in older children and adults; but the frequency of haptoglobin-negatives was not higher in the young children as might have been expected if haemolysis due to malaria were the main cause of haptoglobin depletion. A high incidence of haptoglobin-negatives can occur in a restricted locality, as in the New Guinea highlands where it was found in only one of a number of populations (Barnicot and Kariks, 1960). In some regions, including tropical Africa, with a high incidence of the gene for G-6PD deficiency, haemolytic episodes due to this anomaly may be a contributory factor (Siniscalco, 1959; Allison and Barnicot, 1960). It is unlikely, however, that many of the cases found in western Europeans or in Negroes from the northern United States are due to pathological processes but genetical investigations of families have not entirely clarified the matter. Several pedigrees showing haptoglobin-negative children with a type 2-1 parent exclude the possibility that the former are homozygous for an allele recessive to Hp^1 and Hp^2; nor can a dominant modifier at another locus be involved. Giblett and Steinberg's (1960) data on American Negroes show that this phenotype is more frequent in matings involving 2-1M or Hp-negative parents. They suggest that Hp^{2m} homozygotes or the heterozygotes with Hp^1 or Hp^2 occasionally manifest as haptoglobin-negatives. Cases have been recorded in which no haptoglobins were detectable on a first examination but later specimens showed a weak 2-2 or 2-1 pattern (Sutton *et al.*, 1959; Galatius-Jensen, 1958b; Mäkela *et al.*, 1959).

Distribution

During the five years since the haptoglobin variants were first described, populations from most of the major areas of the world have been examined but the data are still puny in comparison with the voluminous blood group and haemoglobin literature. In dealing with sera collected under unfavourable conditions and sent long distances by air, the possibility of deterioration due to over-long exposure to high ambient temperatures and to bacterial contamination, must not be forgotten. Unusual phenotypes can seldom be checked on repeat specimens or be further investigated in members of the family.

World variations in the frequency of Hp^1 are shown on the map in Fig. 4. Most of the data have recently been summarized by Sutton *et al.* (1960) but some additional values, the sources of which are mentioned below, have been added. In Western European peoples, including the Basques with their serological oddities, the Hp^1 frequency seems to be fairly uniform, ranging around 40 per cent. A rather lower figure (32 per cent) is reported in nomadic Lapps (Beckman and Melbin, 1959) and in the most northerly group examined it was only 28 per cent. The Lapps, it is well known, differ notably from other Europeans in various other genetic traits and also to some extent in morphology. In West and Central Africa, however, we find in the main a much higher range of Hp^1 frequencies. In East Africa the Ganda (Allison and Barnicot, 1960) show a comparably high value but the gene appears to be less frequent in the Masai. Results from Ethiopia (Barnicot *et al.*, 1961), on the other hand, show haptoglobin figures closely similar to those in Europe. The incidence of the Hp^1 gene is also distinctly lower in a number of South African Bantu peoples (Barnicot *et al.*, 1959; Barnicot and Zoutendyk, 1961) than in most Africans of the tropical zone. So far the Bushmen, represented by a sample from Bechuanaland (Barnicot *et al.*, 1959), are outstanding with an Hp^1 frequency of only 29 per cent, and it is interesting to find that a people in whom the African speciality of high R_0 is carried to

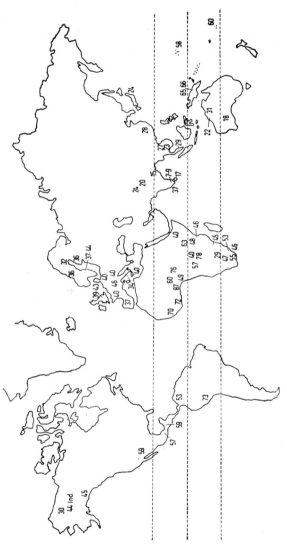

FIG. 4. World distribution of haptoglobin gene Hp[1] (%).

an extreme, show the opposite trend in another genetic trait. Most Asiatic populations so far tested, including a diversity of peoples from Northern and Southern India, Malays, Thai, Japanese and Chinese have much lower Hp^1 frequencies ranging from about 25 per cent to around 10 per cent (Matsunaga and Murai, 1960; Kirk et al., 1960; Kirk and Lai, 1961). In a large sample of aborigines from the Western Desert of Australia, Kirk (1961) has found a frequency of 15·5 per cent, but somewhat higher values in aborigines of the northwest coast region. In the highlands of New Guinea, however (Barnicot and Kariks, 1960; Bennett et al., 1961), the gene reaches an incidence comparable to tropical Africa. The list of peoples with low Hp^1 frequencies includes ones as diverse in appearance as Pathans of the North West Frontier (Kirk and Lai, 1961), Chinese and Australian aborigines, and the distribution evidently bears little or no relation to traditional classifications based on morphological criteria. Information from the New World is still scanty, but Hp^1 is found (Allison et al., 1960) to be fairly low (30 per cent) in Alaskan Eskimos, distinctly higher in Athabascan Indians from this region and higher still in the linguistically related Apache of New Mexico, and in the Maya and other peoples of Central America (Sutton et al., 1958; Sutton, et al., 1960). Arends and Rodriguez (1960) found an Hp^1 frequency of 53 per cent in the Paraujano of western Venezuela, though some admixture with both Negroes and Europeans cannot be excluded. Best and Giblett (1961) also found relatively high Hp^1 frequencies (0·71, 0·74) in two Peruvian Indian groups. In concluding a brief survey of haptoglobin distribution in man it should be mentioned that examples of a variant resembling human 2-1M have been found in a monkey, Macaca irus (Beckman and Cedermark, 1960), but that a single band resembling that of human type 1-1 has been the only pattern detected so far in various other monkey species and also in two chimpanzees (Mäkela et al., 1960; Blumberg, 1960; Baitsch, 1960).

Natural Selection

It is still a matter of conjecture how the haptoglobin polymorphism is maintained and what factors have been responsible for the wide regional differences in gene frequency. By binding free haemoglobin the haptoglobins prevent its passage through the renal glomerulus (Allison, 1959b), thereby conserving iron and averting kidney damage, but in some tropical peoples, as we have seen, sufficient haemolysis to deplete the haptoglobins is quite common. The haemoglobin binding capacity of type 1-1 is on average higher than that of type 2-1 and this in turn higher than that of type 2-2 (Nyman, 1959) so that from this point of view the heterozygote has no obvious advantage over type 1-1. In Africa and perhaps in the New World also Hp^1 seems to be commoner in the tropical zone but there are exceptions to this in Asia. In Liberian tribes a tendency was noted for Hp^1 frequencies to be higher in those with high sickling rates (Sutton *et al.*, 1958) and there is some indication of such a relationship in other parts of Africa. In India, however, the Irula with 30 per cent sickling (Lehmann and Cutbush, 1952) have the conspicuously low Hp^1 figure of 7 per cent (Kirk and Lai, 1961). The haptoglobin concentration in the plasma is known to show fluctuations in various nonhaemolytic diseases (Nyman, 1959), and in a wide variety of inflammatory and infectious conditions the level may be considerably raised. Murray and Connell (1960) recently showed that in rabbits the concentration rose to more than ten times its normal value after subcutaneous injection of turpentine. The significance of these changes is obscure at present but they suggest that haptoglobins may have other functions besides that of haemoglobin carriers. Haptoglobins are generally not detectable by starch-gel methods at birth (Galatius-Jensen, 1957) though traces can be shown immunologically (Fine and Battistini, 1960). The antigenic properties of the three common phenotypes are closely similar at least when tested against animal antisera. It seems unlikely that maternal incompatibility effects exert a selective action and family studies to date do not suggest significant prenatal elimination of particular genotypes, attributable to this kind of interaction.

Transferrins

A major portion of the β-globulin fraction consists of protein which combines firmly with free ferric iron and is known as transferrin or siderophilin. Plasma iron is normally bound to this carrier and toxic symptoms ensue if iron in excess of the binding capacity is injected. Transferrin is considered to act as a physiological iron transporter, liberating it to the tissues, especially the bone marrow (Laurell, 1960).

In starch-gel electrophoresis a well-defined band between the first Hp band and the albumen (Fig. 1) can be identified as transferrin by locating the radioactive regions of the gel after adding Fe^{59} (Allison, 1959a) to the serum or injecting it into the bloodstream; alternatively, purified transferrin can be run in parallel with normal serum.

β-Globulin variants which were later proved to be transferrin were first described by Smithies (1957). In sera from American Negroes and from Australian aborigines he found cases in which in addition to the common transferrin band, denoted TfC, there was a second band with a slower mobility under the standard conditions. This variant is now called TfD_1 to distinguish it from a number of others, described by Smithies, by Giblett and by Harris and their co-workers, which move either slower or faster than TfC and with varying mobilities (Giblett et al., 1959). In the slower, D series, we have, in order of decreasing mobility, TfD_0 found in an American Negro, TfD_1 mentioned above, TfD_2 from a Gambian Negro (Harris et al., 1958a), and TfD_3 again from an American Negro. There is also a series of B variants, TfB_2, TfB_1, TfB_0 which migrate faster than TfC and were first described in subjects of European descent.

Genetical information about these variants is still limited since all of them except TfD_1 have proved to be rather rare in the populations so far examined. At least in the cases of TfD_1, TfB_1 and TfB_2 the evidence indicates that each is determined by a single gene (Horsfall and Smithies, 1958; Smithies and Hiller, 1959). TfD_1 is common in American Negroes, some African peoples and Australian aborigines, so that a number of matings of presumed heterozygotes yielding a homozygous type in which TfD_1 alone is present, have been recorded. It is generally supposed that the transferrin variants depend on a series of alleles at the locus designated Tf; this may well prove to be the case but due to the rarity of some of them and the tendency for certain ones to be restricted to particular populations, families segregating for more than one variant are not easily found and have not been reported.

Since the transferrin variants differ in mobility from one another both in paper electrophoresis and in starch-gel, it has been suggested that the main difference between them is in charge rather than molecular size.

Some frequency figures for TfD_1 have been collected together in Table I. By far the highest frequency value for this genre is found in Australian aborigines with 19·6 per cent. In a later survey in north western Australia, however, considerably lower values, comparable to those in the New Guinea Highlands were found (Kirk, 1961). In peoples from the New Guinea highlands, who, as mentioned above, are very different from the Australians in hapto-globins, the frequency falls to 8·4 per cent while in Borneo it appears that it may be in the region of 5 per cent (Harris et al., 1959b). In the Indian region it was not found in various tribal groups of the south such at Kotas, Irulas, Kurumbas and Todas, though the samples are not large, nor in Tamils or in Punjabis or Pathans. In the Oraon, a tribal group of Bihar, however, the frequency was 3·2 per cent. While the Sinhalese showed only a low frequency (0·6 per cent) the gene was much commoner in the Veddahs (6·3 per cent). A low but appreciable incidence was observed in Thailand (5·0 per cent and 3·1 per cent) in Malays (2·8 per cent) and in Chinese (4·4 per cent).

Turning to Africa we find that the Habe (Hausa) of Northern Nigeria have 7·5 per cent and the figure is probably about the same for the Yoruba in the southwest; in Nigerian Fulani (3·2 per cent) and Gambians (1·0 per cent), on the other hand, the incidence appears to be lower. The Ganda in East Africa showed only 1·5 per cent and TfD_1 was not detected in a small sample of Masai or in Ethiopians. The figure for Bushmen (6·6 per cent) is comparable to the higher range in West Africa and the Bantu-speaking Tswana, also from Bechuanaland, showed a moderately high figure (though very different in Hp frequencies). In certain other Bantu peoples of South Africa the gene was found to be less common.

For the New World aborigines I know only the report of Sutton and his co-workers (1960) for Central America. Only five examples of a CD phenotype were found in a sample of 638, together with four cases of fast (B) variants, and it is possible that these may have resulted from mixture with Negroes or with Europeans.

Other primates have not been neglected. Blumberg (1960) found only a single transferrin, slower than human TfD_1 in a few chim-panzees but a wide variety of patterns in a large sample of rhesus monkeys (*Macaca mulatta*). Goodman and Poulik (1961) have very recently confirmed the occurrence of multiple transferrin phenotypes in this species of monkey; they recognized as many as 14 different patterns, some consisting of single bands of various mobilities, others of double bands and one with three bands. In another Macaque species (*Macaca irus*), however, only one transferrin type was observed. The genetic basis of this remarkable transferrin polymorphism in *M. mulatta* has not been worked out. It is reminiscent of the complex situation in cattle where five alleles

TABLE I
DISTRIBUTION OF TRANSFERRIN TfD₁

Population	N	C	Phenotypes CD₁	D₁	Gene TfD₁ %
Australian aborigines (W. Australia)	349	226	109	14	19·6
New Guinea (Highlands)	518	434	81	3	8·4
Veddahs (Ceylon)	64	57	6	1	6·3
Oraon	125	117	8	0	3·2
Sinhalese	87	86	1	0	0·6
Tamils (Malaya)	133	133	0	0	0·0
Todas	89	89	0	0	0·0
Kotas	11	11	0	0	0·0
Irulas	74	74	0	0	0·0
Kurumbas	49	49	0	0	0·0
Punjabis	207	207	0	0	0·0
Pathans	185[1]	183	0	0	0·0
North Thais	139[2]	124	14	0	5·0
South Thais	274	258	15	1	3·1
Malays	235	224	11	0	2·8
Chinese (Malaya)	103	95	7	1	4·4
Habe (N. Nigeria)	120	102	18	0	7·5
Bushmen	113	99	13	1	6·6
Tswana	152	137	15	0	4·9
Fulani (N. Nigeria)	111	104	7	0	3·2
Shangaan (S.African)	172	162	10	0	2·9
Ganda	165	160	5	0	1·5
Xhosa (S. Africa)	69	67	2	0	1·5
Zulu (S. Africa)	116	113	3	0	1·3
Gambians	153[3]	149	3	0	1·0
Baca (S. Africa)	97	96	1	0	0·5
Ethiopians	312	312	0	0	0·0

[1] Including one B₁C, one B₂C
[2] Including one CD₀
[3] Including one CD₂

controlling β-globulin variation are postulated, each allele controlling a series of four bands. The β-globulin variants in question have been shown to bind iron (Giblett et al., 1959). Comparison of monkey and chimpanzee sera with human by means of two-dimensional starch-gel electrophoresis (Goodman et al., 1960) shows that the pattern of proteins in the chimpanzee is much more like the human than is that of Macaques, though there are some quali-

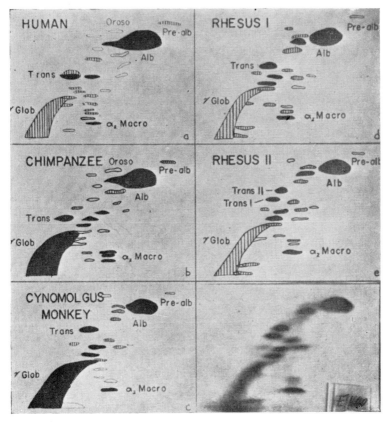

FIG. 5. Diagrammatic serum protein patterns of Man, Macaques and Chimpanzee as seen in two dimensional starch-gel electrophoresis. In addition, a photograph of the serum pattern of the rhesus monkey is presented.
(From Goodman et al., Nature, 188, 78, 1960. By permission of the Editors of Nature.)

tative and quantitative differences (Fig. 5). In the Macaques orosomucoid (prealbumin 2) is absent.

Regarding selective pressures acting on transferrin genes we are still ignorant. Substantial variations in the saturation of transferrins with iron may occur in diseases such as pernicious anaemia and chronic iron deficiency (Laurell, 1960). In the latter condition and also in pregnancy the concentration of transferrin in the plasma is increased. No unusual incidence of particular transferrin variants in particular diseases has so far been found. It appears that the variant forms do not differ from TfC in their iron-binding capacity (Turnbull and Giblett, 1960), but it is still possible that in some circumstances there may be differences in the facility with which they transfer it to the tissues. A quite different line of speculation is suggested if a preliminary report (Martin and Jandl, 1959) claiming that transferrin inhibits multiplication of some viruses in tissue-culture is confirmed. Some curious findings in cattle are thought to indicate prenatal elimination of certain phenotypes according to whether they carry the same or different transferrins to those of the mother (Ashton, 1959a), and human data as it accumulates, should obviously be scrutinized for similar effects. In British cattle a gradient of increasing frequency for the allele β^E in passing from southern to more northerly breeds (Ashton, 1958a) and the relatively high frequency of this gene in zebu cattle (Ashton, 1959b) has given rise to the idea that it is in some way related to adaptability to climatic stress. Although in the Old World, at least, TfD_1 appears to be most frequent in the inhabitants of hot climates, in the present inadequate state of our information further speculation is pointless; but as results for more populations in a wider range of habitats become available, it will certainly be interesting to look for correlations between gene frequencies and various indices of climatic stress.

References

ALLISON, A. C. (1959a) Identification of human serum proteins binding iron, copper and thyroid hormones by starch-gel electrophoresis. *Experientia* **15,** 281.

ALLISON, A. C. (1959b) Genetic control of human haptoglobin synthesis. *Nature (Lond.)* **183,** 1312.

ALLISON, A. C., BLUMBERG, B. S. and AP REES, W. (1958) Haptoglobin types in British, Spanish Basque and Nigerian African populations. *Nature (Lond.)* **181,** 824.

ALLISON, A. C., BLUMBERG, B. S. and GARRY, BARBARA (1960) Haptoglobins and haemoglobins of Alaskan Eskimos and Indians. *Ann. Hum. Genet.* **23**, 349.

ALLISON, A. C. and BARNICOT, N. A. (1960) Haptoglobins and transferrins in some East African peoples. *Acta. Genet.* **10**, 17.

ARENDS, T. and RODRIGUEZ, M. L. G. DE (1960) Haptoglobin types in a Paraujano Indian population. *Vox. Sanguinis* **5**, 452.

ASHTON, G. C. (1958) Genetics of Beta-Globulin polymorphism in British cattle. *Nature (Lond.)* **182**, 370.

ASHTON, G. C. (1959a) β-globulin polymorphism and early foetal mortality in cattle. *Nature (Lond.)* **183**, 404.

ASHTON, G. C. (1959b) β-globulin alleles in some Zebu cattle. *Nature (Lond.)* **184**, 1135.

BAITSCH, H. (1960) Zur Kenntnis der Haptoglobin-Typen einiger Cercopithecinae. *Athrop. Anz.* **24**, 63.

BARNICOT, N. A., GARLICK, J. P., SINGER, R. and WEINER, J. S. (1959) Haptoglobin and transferrin variants in Bushmen and some other South African peoples. *Nature (Lond.)* **184**, 2042.

BARNICOT, N. A., GARLICK, J. P. and ROBERTS, D. F. (1960) Haptoglobin and transferrin inheritance in northern Nigerians. *Ann. Hum. Genet.* **24**, 171.

BARNICOT, N. A. and KARIKS, J. (1960) Haptoglobin and transferrin variants in peoples of the New Guinea Highlands. *Med. J. Aust.* **II**, 859.

BARNICOT, N. A., GARLICK, J. P., ADAM, A. and BAT-MIRIAM, M. (1961) A survey of some genetical characters in Ethiopian tribes. I. Haptoglobins and Transferrins. *Amer. J. Phys. Anthop.* **20**, 171.

BARNICOT, N. A. and ZOUTENDYK, A. (1961) Unpublished data.

BEARN, A. G. and FRANKLIN, E. C. (1959) Comparative studies on the physical characteristics of the heritable haptoglobin groups of human serum. *J. Exp. Med.* **109**, 55.

BECKMAN, L. and MELBIN, T. (1959) Haptoglobin types in the Swedish Lapps. *Acta Genet.* **9**, 306.

BECKMAN, L. and CEDERMARK, G. (1960) Haptoglobin types in Macaca irus. *Acta. Genet.* **10**, 23.

BENNETT, J. H., AURICHT, C. O., GRAY, A. J., KIRK, R. L. and LAI, L. Y. C. (1961) Haptoglobin and transferrin types in the Kuru region of Australian-New Guinea. *Nature (Lond.)* **189**, 68.

BEST, W. and GIBLETT, E. R. (1961) Personal communication.

BLUMBERG, B. S. (1960) Biochemical polymorphism in animals: Haptoglobins and transferrins. *Proc. Soc. exp. Biol. N.Y.* **104**, 25.

CONNELL, G. E. and SMITHIES, O. (1959) Human haptoglobins: estimation and purification. *Biochem. J.* **72**, 115.

FINE, J. M. and BATTISTINI, A. (1960) Etude immunologique des Haptoglobins humaines individuelles. *Experientia* **16**, 57.

GALATIUS-JENSEN, F. (1957) On haptoglobins in relation to age and disease. *Trans. 6th Congress European Soc. Haematol. Copenhagen,* p. 269.

GALATIUS-JENSEN, F. (1958a) On the genetics of the haptoglobins. *Acta. Genet.* **8**, 232.

GALATIUS-JENSEN, F. (1958b) Rare phenotypes in the Hp system. *Acta. Genet.* **8**, 248.

GIBLETT, E. R. (1959) Haptoglobin types in American Negroes. *Nature (Lond.)* **183**, 192.

GIBLETT, E. R., HICKMAN, C. G. and SMITHIES, O. (1959) Serum transferrins. *Nature (Lond.)* **183**, 1589.

GIBLETT, E. R. and STEINBERG, A. G. (1960) The inheritance of serum haptoglobin types in American Negroes: Evidence for a third allele Hp^{2m}. *Amer. J. Hum. Genet.* **12**, 160.

GOODMAN, M., POULIK, E. and POULIK, M. D. (1960) Variations in the serum specificities of higher Primates detected by two dimensional starch-gel electrophoresis. *Nature (Lond.)* **188**, 78.

GOODMAN, M. and POULIK, E. (1961) Effects of specialisation on the serum proteins in the genus Macaca with special reference to the polymorphic state of transferrins. *Nature (Lond.)*. In press.

HARRIS, H., ROBSON, E. B. and SINISCALCO, M. (1958a) β-globulin variants in man. *Nature (Lond.)* **182**, 452.

HARRIS, H., ROBSON, E. B. and SINISCALCO, M. (1958b) A typical segregation of haptoglobin types in man. *Nature (Lond.)* **182**, 1324.

HARRIS, H., ROBSON, E. B. and SINISCALCO, M. (1959a) Distribution of serum haptoglobin types in some Italian populations. *CIBA Symp. Medical Biology and Etruscan origins,* London, Churchill, p. 220.

HARRIS, H., ROBSON, E. B. and SINISCALCO, M. (1959b) Genetics of the plasma protein variants. *CIBA Symp. Biochemistry of human genetics,* London, Churchill, p. 155.

HIRSCHFELD, J. (1959) Immuno-electrophoretic demonstration of qualitative differences in human sera and their relation to haptoglobins. *Acta. Path. Microbiol. Scand.* **47**, 160.

HIRSCHFELD, J. (1960) Immuno-electrophoretic differentiation of hapto-globins from another group-specific inheritable system in normal sera. *Nature (Lond.)* **187**, 126.

HORSFALL, W. R. and SMITHIES, O. (1958) Genetic control of some human serum β-globulins. *Science* **128**, 35.

KIRK, R. L., LAI, L. Y. C., MAHMOOD, SYED and BHAGWAN SING, R. (1960) Haptoglobin types in South-East Asia. *Nature (Lond.)* **185**, 185.

KIRK, R. L. and LAI, L. Y. C. (1961) The distribution of haptoglobin and transferrin groups in South and South-East Asia. *Acta. Genet.* In press.

KIRK, R. L. (1961) Personal communication.

LAURELL, C. B. (1960) Metal-binding plasma proteins and cation transport. *The Plasma Proteins,* Vol. 1, ed. Putnam, p. 349. New York Academic Press.

LEHMANN, H. and CUTBUSH, MARIE. (1952) Sickle-cell trait in Southern India. *Brit. Med. J.* **1**, 404.

MÄKELA, O., ERIKSSON, A. W. and LEHTOVAARA, R. (1959) On the inherit-ance of the serum haptoglobin groups. *Acta. Genet.* **9**, 149.

MÄKELA, O., RENKONEN, O. V. and SALONEN, EEVA. (1960) Electrophoretic patterns of haptoglobins in apes. *Nature (Lond.)* **185**, 852.

MARTIN, C. M. and JANDL, J. H. (1959) Inhibition of virus multiplication by transferrin. *J. Clin. Invest.* **38**, 1024.

MATSUNAGA, E. and MURAI, KYOKO. (1960) Haptoglobin types in a Japanese population. *Nature (Lond.)* **186**, 320.

MURRAY, R. K. and CONNELL, G. E. (1960) Elevation of serum haptoglobin in rabbits in response to experimental inflammation. *Nature (Lond.)* **186**, 86.

NYMAN, M. (1959) Serum haptoglobins: methodological and clinical studies. *Scand. J. Clin. and Lab. Invest.* **2**, Suppl. 39.

POULIK, M. D. (1957) Starch-gel electrophoresis in a discontinuous system of buffers. *Nature (Lond.)* **180**, 1477.

SMITHIES, O. (1955) Zone electrophoresis in starch-gels: group variations in the serum proteins of normal adults. *Biochem. J.* **61**, 629.

SMITHIES, O. (1957) Variations in human serum β-globulins. *Nature (Lond.)* **180**, 1482.

SMITHIES, O. (1959a) An improved procedure for starch-gel electrophoresis: further variations in the serum proteins of normal individuals. *Biochem. J.* **71**, 585.

SMITHIES, O. (1959b) Zone electrophoresis in starch-gels and its application to studies of serum proteins. *Recent Advances in Protein Chemistry* **14**, 65.

SMITHIES, O. and WALKER, N. F. (1956) Genetic control of some serum proteins in normal humans. *Nature (Lond.)* **176**, 1265.

SMITHIES, O. and CONNELL, G. E. (1959) Biochemical aspects of the inherited variations in human haptoglobins and transferrins. *CIBA Symp. Biochemistry of Human Genetics*, p. 178. London, Churchill.

SMITHIES, O. and HILLER, O. (1959) The genetic control of transferrins in humans. *Biochem. J.* **72**, 121.

SINISCALCO, M. (1959) *CIBA Symp. Biochemistry of Human Genetics.* Discussion, p. 175. London, Churchill.

SUTTON, H. E., NEEL, J. V., LIVINGSTONE, F. B., BINSON, G., KUNSTADTER, P. and TROMBLEY, L. E. (1959) The frequencies of haptoglobin types in five populations. *Am. J. Hum. Genet.* **23**, 175.

SUTTON, H. E., MATSON, G. A., ROBINSON, A. R. and KOUCKY, R. W. (1960) Distribution of Haptoglobin, Transferrin and Haemoglobin types among Indians of Southern Mexico and Guatemala. *Amer. J. Hum. Genet.* **12**, 338.

TURNBULL, A. and GIBLETT, E. (1960) The binding and transport of iron by unusual transferrins. *Clin. Res. Proc.* **8**, 133.

URINARY AMINO-ACIDS

Stanley M. Gartler

The position remains as indicated here by Dr. Gartler that no other urinary amino-acids of anthropological interest have been uncovered, that is to say no other variants attaining appreciable frequencies in different populations. However, the use of urinary amino-acid excretion patterns to identify clinical states of genetic origin has made considerable progress, so that those genetic metabolic diseases that show typical population frequency differences are an obvious field awaiting further investigation, and so are the identification of cultural factors (diet, drugs) that affect the levels of excretion.

THE development of chromatographic techniques has led to a widespread exploration of the biochemical contents of various fluids and tissues. During this past decade, human urines, among other materials, have been screened for variations in their content of amino-acids, phenolic compounds, carbohydrates, steroids, fluorescent substances and other compounds. Thousands of urine specimens have been analysed by chromatographic techniques for amino-acids alone. Almost at the very beginning of this period a new naturally occurring amino-acid, β-aminoisobutyric acid (BAIB), was detected by chromatographic means in large amounts in some human urine specimens (Crumpler et al., 1951; Fink, Henderson and Fink, 1951). The studies that soon followed of the variability in excretion rate of BAIB indicated its genetical and anthropological interest. Some ten years and perhaps 100,000 or more chromatograms later, I am sorry to say that no other urinary amino-acids of anthropological interest have been uncovered. For this reason I will begin with, and spend the greater part of my talk on, the genetico-anthropological aspects of β-aminoisobutyric aciduria and conclude with some comments on the prospects for further discoveries of urinary amino-acid variations of anthropological interest.

β-Aminoisobutyric Aciduria

Historical

Dent (1948) was one of the first workers to apply chromatographic techniques to the examination of the amino-acid content of human urine. Among the unknown ninhydrin positive substances found in his early work was one designated as "T" spot, since it was first detected in the urine of a subject named Trotter. A short time later Crumpler *et al.* (1951), and Fink, Henderson and Fink (1951) independently identified this substance as β-aminoisobutyric acid. Since that time BAIB has also been identified in small amounts in a marine organism (*Mytilus edulis*, Awapara and Allen, 1959) and a plant (*Iris tringitana*, Asen *et al.*, 1959), but in no other forms although searches for its presence have been made in other mammalian and primate species (Gartler, Firschein, and Gidaspow, 1956). Thus at present this metabolic polymorphism appears restricted to man.

Detection and Quantification of Urinary BAIB

There are no specific colorimetric or enzymatic methods for the detection of BAIB and consequently it is necessary to isolate the substance before any quantification is possible. Suitable fractionation of mixtures containing BAIB have been achieved by column chromatography, two-dimensional paper chromatography, and unidimensional high voltage paper electrophoresis. Quantification is achieved by determining the intensity of the colour formed by the isolated substance with ninhydrin.

Of the three methods of fractionation, column chromatography (Stein, 1953) has the smallest experimental error and offers the greatest amount of simultaneous additional information on other amino-acids in the material under investigation. However, it is much too slow and costly a method to be of practical value in either genetical or anthropological studies. Two-dimensional paper chromatography (Dent, 1948) has the largest experimental error, but is relatively rapid and has been successfully applied in both genetical and anthropological investigations (Harris, 1953; Sutton and Clark, 1955; and Gartler, Firschein and Gidaspow, 1956). It also offers a considerable amount of simultaneous information on other urinary amino-acids. Unidimensional high voltage paper electrophoresis (Kickhofen and Westphal, 1952) is the most rapid, simple and sensitive of the three fractionation methods. It is somewhat more repeatable than two-dimensional paper chromatography but offers the least additional information on other amino-acids. We have been using this last method in our laboratory for the past three years on genetical, anthropological and metabolic investigations of β-aminoisobutyric aciduria. It it particularly useful for large-scale surveys and for this reason I will describe it in some detail.

The system consists of a power supply and an electrophoresis tank. The high voltage power supply should have a capacity of 2000 to 3000 volts and 0·2 to 0·3 A. The electrophoresis tank has an outer and inner compartment and can be made by simply putting one aquarium jar inside of a slightly larger one. Buffer (pH 3·6, approximately 89 water, 10 acetic acid and 1 pyridine) is put in the bottom of both jars and an immiscible solvent (e.g. toluene) added so as to form a continuous liquid phase between the two compartments. Electrodes from the power supply are placed in the buffer layer of each compartment. The supporting medium (Whatman 3 MM filter paper) must be long enough to extend between the buffers in the bottom of each compartment.

Untreated urine specimens (1 to 2 μl) are applied to the filter paper with a micro pipette and allowed to dry. Specimens are spotted 2 cm apart on the sheet and placed so that when the sheet is suspended in the tank the spots will be in the toluene phase. After applying the samples, the paper is wetted with the buffer and the excess blotted off. It is important not to apply buffer directly to the spotted area but rather to let the buffer diffuse into it from other wetted areas. The sheet is then suspended in the tank between the two buffers, and the current applied for 35 min. (approximately 30 V/cm length and 4 ma/cm width). At the end of the run the sheet is removed and air dried overnight. After drying, the sheet is developed by dipping it in a ninhydrin solution (0·2 per cent ninhydrin in 90 per cent isopropyl alcohol) and heating at 105°C for 8 min. Colour intensity is then determined by densitometry. In Fig. 1 is shown a typical run illustrating separation of BAIB from the other urinary amino-acids. β-alanine overlaps

BAIB somewhat in this system, but since it rarely occurs in urine and since it can be distinguished from BAIB, this is not considered a serious problem.

Either toluene or thymol is added to specimens for preservation in transit, and they are stored in the deep freeze at the laboratory. Under these conditions samples have been kept for a year without significant deterioration of BAIB.

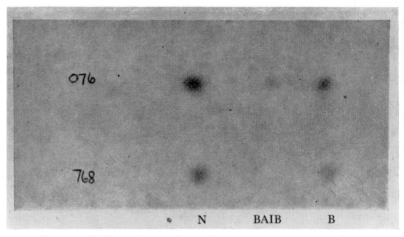

Fig. 1. Unidimensional electrophoretic separation of urinary amino acids at pH 3·6. N = neutrals; BAIB = β-aminoisobutyric acid; B = basic amino acids; 768, low excretor; 076, high excretor.

Ideally excretion rates should be expressed in terms of amount excreted per unit time. This requires careful collection of timed specimens which is essentially impossible in the field, and not practical for large genetic studies. An alternative to this method is to express the urinary BAIB concentration relative to the concentration of another urinary component, preferably one with a constant excretion rate. Creatinine has been the usual choice for this purpose and for fresh specimens it is quite adequate. However, when samples have been standing for some time in transit or even in the laboratory, and especially if the urine is alkaline, a considerable amount of the creatinine is converted to creatine. In such cases, spuriously high values of BAIB excretion may result, and consequently it is safest when using this method to express BAIB concentration relative to total creatinine concentration (any creatine present is converted to creatinine before determination).

A simpler method and one with a lower experimental error is to express BAIB concentration relative to that of another amino-acid determined on the same sheet. This method has been used successfully in several studies (Harris, 1953; Calchi-Novati *et al.*, 1954; Gartler, Firschein and Kraus, 1957) employing either alanine or glycine as the other amino-acid. Though the excretion rates of these and other amino-acids are not as constant as creatinine, satisfactory discrimination between different BAIB excretion rates has been achieved. In the unidimensional electrophoretic separation described here glycine, alanine, and the other neutral amino-acids are not separated from one another. Consequently, the entire neutral fraction is used as the basis for expressing BAIB concentration (i.e. optical density BAIB spot/optical density neutral amino-acid spot). It is important that both the neutral and BAIB spots be in their linear range with regard to optical density. This requirement is usually satisfied with one to two μl quantities of urine, but smaller amounts may be required in concentrated samples. Duplicate determinations by this method had errors of under 10 per cent. It should be mentioned that satisfactory separation of glycine and alanine, but not BAIB may be achieved by running the electrophoretic separation at pH 2·2 (Visakorpi and Puranen, 1958).

Genetics of BAIB Excretion

A series of genetic studies utilizing a variety of approaches have all demonstrated that most of the variation between individuals in their excretion rates of BAIB is under genetic control (Harris, 1953; Sutton and Vandenberg, 1953; Calchi-Novati *et al.*, 1954; Gartler, Dolzhansky and Berry, 1955; Gartler, 1956; Gartler, Firschein and Kraus, 1957; and Grouchy and Sutton, 1957). In the studies where individuals were classified as either high or low excretors, the resulting data were compatible with a simple monogenic determination of the observed variability, high excretors being homozygous for a single recessive gene and low excretors either heterozygous or homozygous for the dominant allele (Harris, 1953; Calchi-Novati *et al.*, 1954; Gartler, 1956; and Gartler, Firschein and Kraus, 1957). Such classification is somewhat arbitrary since the distribution of the variability in BAIB excretion rates is a continuous one, and so it was clear that other factors than three genotypes must be contributing to the observed variability. The question then arose as to whether these "other factors" were a combination of environmental, technical and possibly minor genetic ones or that this was not a simple genetic system but rather one under multifactorial genetic control.

The most important evidence for distinguishing between these possibilities is the nature of the distribution. If the genetic system involved is simple with "other modifying factors" then a bimodal continuous distribution is expected, whereas with multifactorial genetic control or a complicated multiple allelic system, a unimodal normal distribution is expected. Evidence for bimodality has been obtained (Gartler, Firschein and Kraus, 1957) and in Fig. 2 such a distribution is shown. There is considerable overlap but considering the magnitude of environmental variability and technical errors this is perhaps not unexpected. Normal day to day differences in BAIB excretion rates for an individual exceed 25 per cent (Gartler, 1959) and in most of the genetic studies technical errors were over 10 per cent. It might be mentioned at this point that there are also

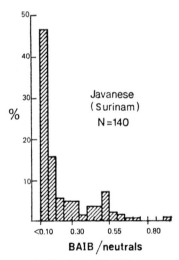

FIG. 2. Frequency distribution of BAIB excretion (optical density BAIB spot/optical density neutral amino acids spot) in a Javanese population.

several specific non-genetic factors which can markedly increase BAIB excretion (exposure to radiation or nitrogen mustard, Rubini *et al.*, 1959, and Awapara and Shullenberger, 1958; starvation, Pare and Sandler, 1957), but these are all extremely abnormal situations and are not pertinent to the above genetic studies. On the other hand, none of the published data are normally distributed. Furthermore, a survey of all the published family data does not suggest any specific complicated genetic mechanism with which they are compatible. In summary, then, the simplest genetic hypothesis with which the data are compatible is that the major variation in BAIB excretion is under the control of a single pair of alleles.

Metabolic Basis of BAIB Excretion

Crumpler *et al.* (1951) postulated that the physiological difference between high and low excretors was a renal tubular defect in the reabsorption of BAIB. However, recent experimental investigations of this hypothesis (Gartler, 1959a; Jagenberg, 1959), have failed to reveal any renal difference in the handling of BAIB by high and low excretors. In fact the evidence strongly suggests that no tubular reabsorption of BAIB occurs either in high or low excretors. By elimination, then, it appears that a difference in intermediary metabolism is the basis of the variation in BAIB excretion.

In Fig. 3 are shown the known metabolic pathways leading to BAIB formation as derived from experimental work with the rat (Fink, Fink and Henderson, 1953, 1956) and the pig (Coon, 1955; Robinson *et al.*, 1957; and Kupiecki and Coon, 1957). The thymine pathway leading to BAIB has been shown to be operative in man (Awapara and Shullenberger, 1957; Gartler, 1959a) but as yet no evidence has been brought forward for the importance of valine as a precursor of urinary BAIB in man. Feeding experiments in our laboratory with up to 5 gm. of DL valine have failed to increase urinary BAIB levels significantly in either high or low excretors. Of further importance in this respect is the "maple syrup urine disease" (Westall *et al.*, 1958) in which among other things there is a markedly elevated plasma level of valine with spillage of valine into the urine, but no apparent effect on BAIB excretion. A comparative study in high and low excretors of the metabolic steps involved in the reductive catabolism of thymine to BAIB has indicated that no metabolic difference is present in the thymine $\rightleftharpoons$ dihydrothymine $\rightleftharpoons$ β-ureidoisobutyric acid $\rightarrow$ BAIB sequence, but that a difference exists in the ability of high and low excretors to metabolize BAIB (Gartler, 1959b). This suggests that the

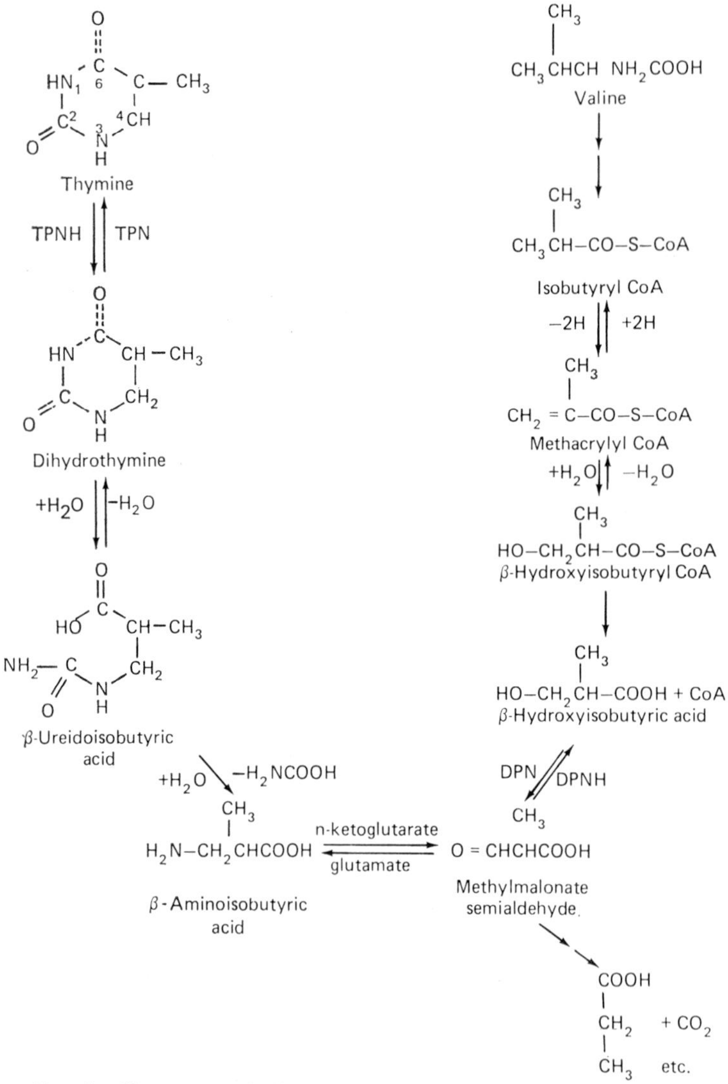

Fig. 3. Known metabolic pathways leading to BAIB formation.

metabolic block is subsequent to BAIB formation possibly at the
BAIB $\rightleftharpoons$ methylmalonate semialdehyde step or in the conversion
of the latter to propionate.

The importance of the reductive catabolism of thymine in the
production of BAIB appears to explain most of the non-genetic
causes of increased BAIB excretion. Increases in tissue destruction
(radiation and nitrogen mustard treatment, degenerative diseases)
or interference with new tissue synthesis (starvation) should lead
to an increase in nucleic acid precursors, including thymine, and its
subsequent conversion to BAIB and excretion. Since even genetic
low excretors have a limited capacity to metabolize BAIB (Gartler,
1959a), they may become high excretors if their thymine level is
sufficiently increased.

Thus genetic high excretion of BAIB appears to result from the
relative inability of high excretors to metabolize BAIB formed
through normal endogenous catabolism of thymine; while non-
genetic induced excretion of BAIB may result from excessive
thymine production due to increased cellular destruction or interference with new cellular synthesis.

Racial Variation in BAIB Excretion

In Table I are given the available distributional data on BAIB
excretion, with individuals classified as either high or low excretors
depending on the apparent position of the antimode, or on the
particular author's classification. It should be pointed out that both
chromatographic and electrophoretic techniques have been used in
the analyses of the reported material. Further, in some of the
studies BAIB concentration was expressed relative to creatinine,
while in others another amino-acid or acids was used as the reference substance. On the positive side it can be stated that studies
of very similar populations (e.g. American whites, Gartler, 1956;
and Sutton, 1960) using different methods have given similar
results. Thus, though the data are not strictly comparable, I believe
broad racial comparisons of the type shown are valid. It might
also be mentioned that there is no evidence to indicate that any
of the non-genetic causes of increased BAIB excretion play a role
in these populational differences.

The most consistent finding from the distributional data is the
low level of high excretion among Caucasoid groups. The Asiatic

Indian sample of Sutton's is most interesting in this respect even though it is a very small one.

On the other hand, in the Negroid and Mongoloid populations sampled the frequencies of high excretors are generally much higher and somewhat more variable. The high values for Mongoloid populations have been substantiated by different investigators using different methods, and most important, through the investigation of various groups (Chinese, Japanese, Thai) living in America (Sutton, 1960).

Table I

BAIB Excretion Rates in Various Populations

Population	% High Excretors	No. of Persons	Reference
Caucasoid			
American	9	504	Average of Sutton and Gartler
English	10	345	Harris
Italian	7	792	Calchi-Novati *et al.*
Asiatic	0	16	Sutton
Negroid			
American	17	63	Average of Sutton and Gartler
Black Caribs	32	285	Gartler, Firschein and Kraus
East Africans	30	40	Gartler, Blumberg and Allison
Mongoloid			
Red Caribs	23	39	Gartler, Blumberg and Allison
American Indians	50	270	Gartler, Firschein and Kraus
Javanese (Surinam)	22	140	Gartler, Blumberg and Allison
Chinese	45	33	Sutton
Japanese	42	41	Sutton
Thai	46	13	Sutton
Micronesians (Marshall Is.)	56	34	Gartler, Blumberg and Allison
Polynesians (Tahiti)	12	46	Gartler, Blumberg and Allison

The high values for the Micronesians (from 56 per cent to over 80 per cent high excretors in two studies, Blumberg and Gartler, 1959; and Gartler, Blumberg and Allison, 1960), would appear to indicate strong Mongoloid affinities for this group. On the other hand, the low level of high excretors among the Polynesians studied is comparable to Caucasoid values.

Though the data are quite limited, it seems clear that we are dealing here with another of the ubiquitous polymorphisms. In such instances it is doubtful whether a great deal more distributional data would be of value in elucidating the population dynamics of the polymorphism. However, the character appears to have considerable anthropological value, and for this purpose further data on racial variation would be most useful.

Evolutionary Aspects of β-Aminoisobutyric Aciduria

Thus far β-aminoisobutyric aciduria has only been found in man. The reason for this apparently restricted species distribution appears explicable from what is known of the comparative biochemistry of BAIB formation. The reductive pathway of thymine catabolism leading to BAIB formation has been examined in three species, the mouse, rat and man (Fink, Fink and Henderson, 1953; Fink et al., 1956; Gartler, 1959a). In the mouse and the rat only a small percentage of administered thymine is converted to BAIB, while both dihydrothymine and β-ureidoisobutyric acid are very effectively converted to BAIB. In man thymine as well as dihydrothymine and β-ureidoisobutyric acid are effectively converted to BAIB. Thus the interspecific difference involves the thymine $\rightleftharpoons$ dihydrothymine step while the intraspecific difference (man) appears to involve a step subsequent to BAIB formation, e.g. BAIB $\rightleftharpoons$ methyl malonate semialdehyde. Such an intraspecific genetic difference in the mouse or rat would not be detectable since the inefficiency of the thymine $\rightleftharpoons$ dihydrothymine step would prevent increased BAIB formation. In essence, then, a double mutant would be required for β-aminoisobutyric aciduria in these lower animals. It may be worth pointing out that we have here an example at the biochemical level of how an interspecific change permits the expression or production of apparently new changes at the intraspecific level.

At present we have no evidence as to the nature of the evolutionary mechanism or mechanisms involved in the development of the intraspecific genetic differences controlling β-aminoisobutyric aciduria. The distributional data indicate a ubiquitous polymorphism. The interracial variation is considerable but does not offer any insight into the population dynamics of this variation. The intraracial variation seems relatively small (Caucasoids especially) and if this should be well documented it could be considered as evidence for the absence, at least at the present, of exogenous selective forces acting on this polymorphism.

A number of reports have appeared indicating possible relationships between β-aminoisobutyric aciduria and various diseases (leukaemia, tuberculosis, diabetes, mongolism, and march hemoglobinuria, see Gartler, 1960, for review). However, in all substantiated cases there is evidence indicating that the association is a secondary one (i.e. increased BAIB excretion from non-genetic causes).

Though our physiological and biochemical knowledge of this polymorphism is far from complete it is already clear that in β-aminoisobutyric aciduria we are dealing with a genetic difference affecting terminal steps in the catabolism of thymine. Theoretically such a change would not be expected to have serious deleterious effects and in fact both high and low excretors appear perfectly normal. There should be a slight elevation in plasma BAIB in high excretors. If BAIB has any biological activity, then even a very small difference in concentration would be significant. In this respect it is worth pointing out the observation of Lindan (1954) that BAIB was inhibitory to the growth of yeast. Along this same line one might consider the possibility of testing BAIB for neurogenic activity, since it bears some structural similarity to γ-aminobutyric acid.

In summary, the prospects for unravelling the population genetics of β-aminoisobutyric aciduria are poor. If this is a currently selectively balanced polymorphism, then we must find a biological clue (e.g. relation to disease) to its nature. The failure to find such a clue could indicate that we have simply not exhausted the possibilities, or that significant selective forces are not currently acting on this polymorphism. In polymorphisms such as these, where segregation of phenotypes is hazy, proof of the absence of selective forces or any other critical evolutionary mechanism (e.g. altered segregation ratio) by direct study is probably not possible.

Other Amino-Acid Variants

As I mentioned in the introduction, there have been no other urinary amino-acid variants found which one could class as being of anthropological interest. Considering the fact that BAIB is a non-protein occurring amino-acid, primarily related to pyrimidine catabolism, the previous statement may be considered by some to exclude amino-acids completely. This is not to say that there are no known genetic variants affecting amino-acid metabolism and excretion. There are clear-cut variants, some well studied genetically, which involve amino-acids or closely related substances (e.g. cystinuria, glycinuria, Hartnup disease, Fanconi disease, cystathionuria, argino-succinic aminoaciduria, maple syrup urine disease, cystinosis, phenylketonuria, tyrosinosis, alkaptonuria, albinism—see Harris, 1959, for review). These are all rare conditions affecting urinary amino-acid concentration directly through interference with renal function (e.g. cystinuria) or indirectly through a block in the intermediary metabolism of a particular amino-acid (e.g. phenylketonuria). In general, they are inherited as recessives with the homozygous recessive genotype usually being severely affected and the heterozygotes all normal. The normal amino-acid concentration in the urine or plasma is markedly changed in the affected individuals (twenty to one hundred fold), while the heterozygotes may exhibit only slight changes or none at all. The absence of polymorphisms may not be unexpected in these conditions, but what must be considered is that we can only detect reliably those genetic changes affecting metabolism which have very marked effects. There is evidence that normal variations in urinary excretion of a number of amino-acids may be under partial genetic control (Sutton and Vandenberg, 1953; Gartler, Dobzhansky and Berry, 1955; Sutton and Clark, 1955; and Tashian and Gartler, 1957) but the chance of successfully using such small variations for exacting genetic studies or anthropological purposes is almost nil. It must be remembered that we are dealing with variations between individuals in metabolic processes in which there is also considerable intra-individual variability. Consequently, the basic genetic change must have a relatively large phenotypic effect before it can be regularly recognized. If we could measure qualitative aspects of the enzyme protein itself then we could predict almost with certainty that polymorphisms would be found, but whose metabolic effects might be very slight indeed.

Prospects

It should be clear that I do not foresee a rosy future for the discovery of anthropologically interesting variants in urinary amino-acid excretion. I would carry this over also to other essential substances not only in urine but in plasma as well. It would appear that the genetic damage to an enzyme necessary to produce a sufficiently recognizable alteration in metabolism is too great in general to be useful evolutionary material. Though the door may be temporarily closed on amino-acids, I believe there may still be many interesting discoveries in the general field of metabolic polymorphisms. The metabolic map of man becomes better known every day and at the same time, the tools with which we measure various aspects of metabolic activity improve. It should be possible today to pick out the weaknesses, so to speak, in man's metabolic armour—that is, the "β-aminoisobutyric acidurias", find the tools to measure these substances and almost predictably discover new metabolic polymorphisms.

Acknowledgements

The author is indebted to Mr. Richard Nanka-Bruce and Mr. Olgerts R. Pavloskis for expert technical assistance during the course of this work. The work has been supported in part by a grant from the National Science Foundation [G.6187].

References

ASEN, S., THOMPSON, J. F., MORRIS, C. J. and IRREVERE, J. (1959) Isolation of β-aminoisobutyric acid from bulbs of *Iris tingitana* var. Wedgewood. *J. Biol. Chem.* **234**, 343.

AWAPARA, J. and ALLEN, K. (1959) Occurrence of β-aminoisobutyric acid in *Mytilus edulis. Science* **130**, 1250.

AWAPARA, J. and SHULLENBERGER, C. C. (1957) Urinary excretion of β-aminoisobutyric acid after administration of thymine and nitrogen mustard. *Clin. Chim. Acta* **2**, 199.

BLUMBERG, B. S. and GARTLER, S. M. (1959) High prevalence of high level BAIB excretors in Micronesians. *Nature (Lond.)* **184**, 1990.

CALCHI-NOVATI, C., CEPPELLINI, R., BIANCHO, I., SILVESTRONI, E. and HARRIS, H. (1954) β-aminoisobutyric acid excretion in urine. A family study in an Italian population. *Ann. Eugen. (Lond.)* **18**, 335.

COON, M. J. (1955) Enzymatic synthesis of branched chain acids from amino-acids. *Fed. Proc.* **14**, 762.

CRUMPLER, H. R., DENT, C. E., HARRIS, H. and WESTALL, R. G. (1951) β-aminoisobutyric acid (α-methyl-β-alanine): a new amino-acid obtained from human urine. *Nature (Lond.)* **167**, 307.

DENT, C. E. (1948) A study of the behaviour of some sixty amino-acids and other ninhydrin reacting substances on phenol collidine filter paper chromatograms, with notes on the occurrence of some of them in biological fluids. *Biochem. J.* **43**, 168.

FINK, K., HENDERSON, R. B. and FINK, R. M. (1951) β-aminoisobutyric acid; a possible factor in pyrimidine metabolism. *Proc. Soc. Exp. Biol. N.Y.* **78**, 135.

FINK, R. M., FINK, K. and HENDERSON, R. B. (1953) β-amino-acid formation by tissue slices incubated with pyrimidines. *J. Biol. Chem.* **201**, 349.

FINK, R. M., McGAUGHEY, C., CLINE, R. C. and FINK, K. (1956) Metabolism of intermediate pyrimidine reduction products *in vitro. J. Biol. Chem.* **218**, 1.

GARTLER, S. M. (1956) A family study of urinary β-aminoisobutyric acid excretion. *Amer. J. Hum. Genet.* **8**, 120.

GARTLER, S. M. (1959a) A metabolic investigation of urinary β-aminoisobutyric acid excretion in man. *Arch. Biochem. Biophys.* **80**, 400.

GARTLER, S. M. (1959b) An investigation into the biochemical genetics of β-aminoisobutyric aciduria. *Amer. J. Hum. Genet.* **11**, 257.

GARTLER, S. M. (1960) In *Proc. Conf. Genetic Polymorphisms and Geographic Variations in Disease,* ed. B. S. BLUMBERG. Grune and Stratton, N. Y. 192–213.

GARTLER, S. M., DOBZHANSKY, TH. and BERRY, H. K. (1955) Chromatographic studies on urinary excretion patterns in monozygotic and dizygotic twins. II. Heritability of the excretion rates of certain substances. *Amer. J. Hum. Genet.* **7**, 108.

GARTLER, S. M., FIRSCHEIN, I. L. and GIDASPOW, T. (1956) Some genetical and anthropological considerations of urinary β-aminoisobutyric acid excretion. *Acta Genet.* **6**, 435.

GARTLER, S. M., FIRSCHEIN, I. L. and KRAUS, B. S. (1957) An investigation into the genetics and racial variation of BAIB excretion. *Amer. J. Hum. Genet.* **9**, 200.

GARTLER, S. M., BLUMBERG, B. S. and ALLISON, A. C. (Unpublished results.)

GROUCHY, J. and SUTTON, H. E. (1957) A genetic study of β-aminoisobutyric acid excretion. *Amer. J. Hum. Genet.* **9**, 76.

HARRIS, H. (1953) Family Studies on the urinary excretion of β-aminoisobutyric acid. *Ann. Eugen. (Lond.)* **18**, 43.

HARRIS, H. (1959) *Human Biochemical Genetics,* Cambridge, University Press.

JAGENBURG, O. R. (1959) The urinary excretion of free amino-acids and other amino compounds by the human. *Scand. J. Clin. Lab. Invest.* **11**, Supp. 43.

KICKHOFEN, B. and WESTPHAL, O. (1952) Paper electrophoresis at high potentials for separation of peptides. *Z. Naturf.* **76**, 655.

KUPIECKI, F. P. and COON, M. J. (1957) The enzymatic synthesis of β-aminoisobutyrate. A product of valine metabolism, and of β-alanine, a product of β-hydroxypropionate metabolism. *J. Biol. Chem.* **229**, 743.

LINDAN, R. (1954) The inhibitory effect of β-aminoisobutyric acid on the growth of yeast. *Biochem. J.* **75**, 31.

PARE, C. M. B. and SANDLER, M. (1954) Amino-aciduria in march haemoglobinuria. *Lancet* **266**, 702.

RUBINI, J. R., CRONKITE, C. P., BOND, V. P. and FLIENDER, T. M. (1959) Urinary excretion of β-aminoisobutyric acid in irradiated human beings. *Proc. Soc. Exp. Biol. N.Y.* **100**, 130.

ROBINSON, W. G., NAGLE, R., BACHHAWAT, B. K., KUPIECKI, F. P. and COON, M. J. (1957) Coenzyme A thiol esters of isobutyric, methacrylic, and β-hydroxyisobutyric acids as intermediates in the enzymatic degradation of valine. *J. Biol. Chem.* **224**, 1.

STEIN, W. H. (1953) A chromatographic investigation of the amino-acid constituents of normal urine. *J. Biol. Chem.* **201**, 45.

SUTTON, H. E. (1960) β-aminoisobutyric-aciduria, in *The Metabolic Basis of Inherited Disease,* Ed. STANBURY, J. B., WYNGAARDEN, J. B. and FREDERICKSON, D. S. New York. McGraw-Hill.

SUTTON, H. E. and CLARK, P. J. (1955) A biochemical study of Chinese and Caucasoids. *Amer. J. Phys. Antrop.* n.s. **13,** 53.

SUTTON, H. E. and VANDENBERG, S. G. (1953) Studies on the variability of human excretion patterns. *Hum. Biol.* **25,** 318.

TASHIAN, R. E. and GARTLER, S. M. (1957) Genetic implications of certain physiological processes affecting the metabolism of L-phenylalanine in man. *Amer. J. Hum. Genet.* **9,** 208.

VISAKORPI, J. K. and PURANEN, A. L. (1958) High voltage paper electro-phoresis. A rapid method of determination of urinary amino-acids. *Scand. J. Clin. Lab. Invest.* **10,** 196.

WESTALL, R. G., DANCIS, J., MILLER, S. and LEPITZ, M. (1958) Maple sugar urine disease. *Fed. Proc.* **17,** 334.

DERMATOGLYPHIC PATTERNS

SARAH B. HOLT

The minutiae of dermatoglyphic detail continue to attract much attention. Perhaps the major advances since this paper was written have consisted in the identification of the relationship of the ridge count to the numbers of the sex chromosomes (Penrose 1967, 1968), and of the pronounced dermatoglyphic variation that occurs in the presence of some chromosomal variation such as occurs in, for example, Turner's syndrome and Down's syndrome. These have prompted searches for dermatoglyphic associations of other diseases with varying degrees of genetic contribution to their etiology, on the whole with not a great deal of success (e.g. reviews by Elbualy and Schindeler 1971; Alter 1967) and the problem has emerged of the prognostic significance of those associations that are discovered (Fuller 1973). As regards normal populations, the flow of descriptive studies continues so that the general outlines of the world distribution of dermatoglyphic variation are emerging; particularly interesting is the suggestion of the presence of local dermatoglyphic variation in settled populations.

SINCE the pioneer researches of Sir Francis Galton dermal ridge arrangements have been used to a considerable extent in physical anthropology. There are various reasons for this, two of the most important being that (1) dermatoglyphic characters, unlike most human traits, are not affected morphologically by age and (2) from before birth they are not affected by environment.

Dermal ridge differentiation takes place in the third and fourth months of foetal life. By the end of the fourth month the ridges and their arrangements are in their complete and permanent form. From this time until after death there is no morphological change either in the detailed structure of the ridges or in the patterns formed by them. The only changes are in size, the growth of the ridges keeping pace with the growth of the hands and feet. The patterns formed by dermal ridges are very variable in size, shape and detailed structure. They can, however, be classified into several main types. Moreover, dermatoglyphic features show racial variation.

Prints of ridged skin on fingers and palms can be taken with little difficulty and supply permanent records of the areas concerned. Good impressions of the soles, particularly of adults, are often more difficult to obtain, while the printing of complete toe patterns presents certain technical difficulties. Consequently, finger- and palm-prints have been used more extensively than sole- and toe-prints. Finger-print data from numerous racial samples are comparatively abundant, palm-print data less so, while those from soles are more limited. Data from toe-prints are very meagre.

Early Work on Racial Samples

Galton (1892) compared the finger-prints of five racial series, English, Welsh, Jews, Negroes and Basques. To his disappointment, he could find no peculiar pattern that was characteristic of any of these peoples. The differences between the groups were limited to differences in frequency of the same pattern types. For example, he found that the frequency of arches on index fingers was 13·6 per cent in the English sample, but only 7·9 per cent in the Jewish. Galton summarized his findings: "The only differences so far observed, are statistical, and cannot be determined except through patience and caution, and by discussing large groups."

Wilder (1904) investigated palm- and sole-prints of Maya Indians, Whites, Negroes and Chinese. He also came to the conclusion that there are no dermatoglyphic peculiarities diagnostic of race in the individual, racial differences showing only in the statistical trends.

Racial Variability in Different Dermatoglyphic Areas

Fingers

The main types of finger-print pattern, in order of increasing complexity, are arches, loops and whorls.

This classification depends on the number of triradii present. An arch has no triradius, a loop one and a whorl typically two triradii (see Fig. 1).

Two methods have been used to compare the frequencies of finger patterns in different populations:

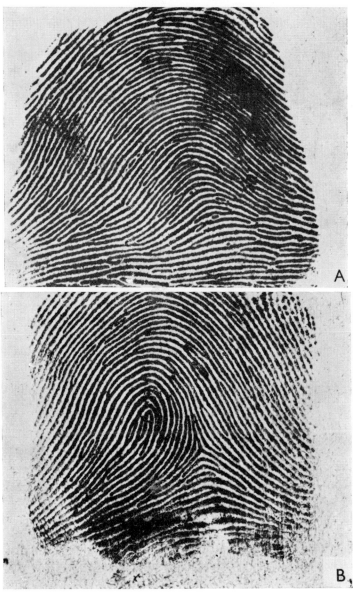

FIG. 1. Finger prints showing the three main pattern-types.
A. Arch (no triradius); B. Loop (1 triradius); C. Whorl (2 triradii)
(*see overleaf*).

(1) Dankmeijer's (1934, 1938) arch/whorl index,

(2) Pattern intensity index.

There is a reciprocal relationship between the frequencies of whorls and arches, a rise in the frequency of one generally being accompanied by a fall in the frequency of the other type. Dankmeijer, therefore, used

$$\frac{\text{the total frequency of arches}}{\text{the total frequency of whorls}} \times 100$$

as an index for population comparisons.

The pattern intensity index is the mean number of triradii found on fingers per individual. This value can be estimated for any racial sample from the frequencies of whorls and loops in the sample, bearing in mind that each whorl has two triradii and each loop one triradius. When the pattern frequencies are given in the form of percentages it is necessary to divide by ten. For individuals, values of the pattern intensity index range from 0 (all arches) to 20 (all whorls). The intensity index can also be used for palms, soles and toes. In the case of palms and soles the triradii have to be counted.

Some examples of the frequencies with which arches and whorls occur on fingers in different races (based on data in the literature) are shown in Fig. 2. The frequencies are given as percentages. The value of the pattern intensity index in each case can be read from the vertical scale. Cummins and Midlo (1943) give a similar graphical representation of data relating to nearly 50 different racial groups.

A recent analysis of this kind by Sachs and Bat-Miriam (1957) concerns the distribution of finger-patterns in Jews who have gone

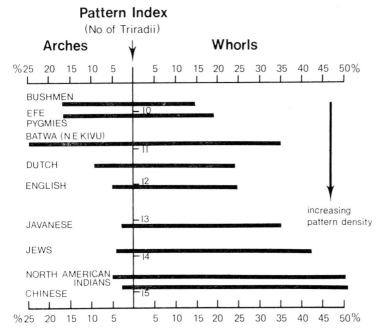

FIG. 2. Frequencies of arches and whorls on fingers in some racial samples. The arrangement is in order of increasing pattern intensity. (Based on data from Cummins and Midlo, 1943, where references to the relevant literature are given.)

to Israel from Bulgaria, Egypt, Germany, Iraq, Morocco, Poland, Turkey and Yemen. These groups were chosen as representatives of various historical migrations. Frequencies of whorls, loops and arches were found in samples of 500 males from each group. Pattern indices varied from 13·30 to 13·98, showing remarkable similarity in a people which has been widely dispersed for centuries in different parts of the world.

Palms

Pattern frequencies are generally used for racial comparison of palmar dermatoglyphics, though main-lines and their terminations have also been studied in some surveys. There are five pattern areas of the palm, the thenar/first interdigital, the second, third and fourth interdigital and the hypothernar areas. The principal pattern-types in the hypothenar area are loops and whorls, but double-looped, S-shaped patterns also occur. Interdigital patterns generally consist of loops opening into the nearest interdigital interval, though small whorls occur very rarely in these areas. Thenar/first interdigital patterns differ from patterns in other areas. They normally have a characteristic appearance with at least some of the ridges running at right angles to the general ridge direction. Most workers have found the frequencies of pattern types occurring in the five dermatoglyphic areas of the palm of a particular race. The frequencies of total patterns occurring in each of these areas are shown for ten races in Fig. 3.

Soles

On the sole there are eight configurational areas, the proximal thenar, the distal and proximal hypothenar, the hallucal (made up of the distal thenar and the first interdigital combined), the second, third and fourth interdigital areas and calcar (heel) area. The region of greatest pattern intensity is the ball of the foot. Calcar patterns are rare. Analyses of sole-prints have usually been limited to the frequencies of pattern-types in the different areas. Many workers have only considered the four areas of the ball region, i.e. the hallucal and the three interdigital areas. Variations in frequencies of total pattern in these four distal regions of the sole are shown diagrammatically in Fig. 4 for eight races.

Wichmann (1956) studied the frequencies of pattern-types in five different areas of the sole in a large series of Germans. He divided

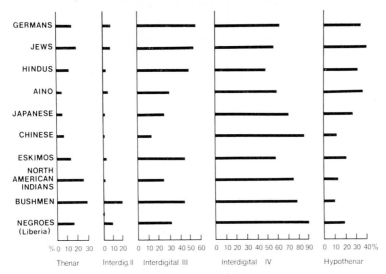

frequency of palmar patterns

FIG. 3. Frequencies of patterns in the five configurational areas of the palm in ten racial samples. (Slightly changed from Cummins and Midlo, 1943, where references to the relevant literature are given.)

the distal sole into three regions, the other two areas being the thenar and the hypothenar. His results are not, therefore, strictly comparable with those of most other workers.

Toes

There is little data on toe-print patterns and what there is concerns four races only. The most extensive series is that of Takeya (1933) for 1000 Chinese. Whorls are less frequent than in fingers, while arches are more frequent. The frequencies of these two patterns in Germans, European-Americans, Chinese and Japanese are given in Fig. 5. The corresponding frequencies of whorls and arches in fingers are also shown.

Trends in Pattern Intensity in White, Yellow-Brown and Black Races

From information of this kind certain general trends emerge. Thus, among White races Nordic populations have a lower finger pattern intensity than European Mediterraneans, while Hindus, Arabs and Jews show somewhat higher intensities than European

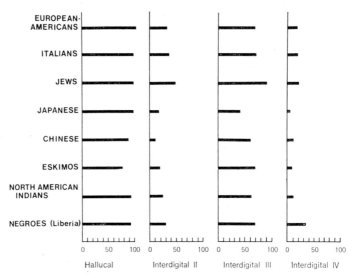

FIG. 4.   Pattern frequencies in the four configurational areas of the distal sole in eight racial samples. (Adapted from Cummins and Midlo, 1943, where references to the relevant literature are given.)

Mediterraneans. Yellow-brown races are characterized by a high pattern intensity in fingers, while in palms there is a reduction of hypothenar, second and third interdigital patterns. There is also reduced pattern intensity in all the four configurational areas of the distal sole. In Black races the pattern intensity of fingers is very variable. Bushmen have the lowest pattern intensity known, while the value for Melanesians is very high. Negroes have intermediate pattern intensities. Palmar hypothenar patterns are commoner than in Yellow-brown races, but less frequent than in Whites. Second and fourth interdigital patterns tend to be numerous, while on soles hallucal and fourth interdigital patterns are common.

Sex Differences

Sex has not been taken into consideration in any of the examples given. There are, however, sex differences within races. For example, on fingers females show almost without exception a higher frequency of arches than males and usually fewer whorls.

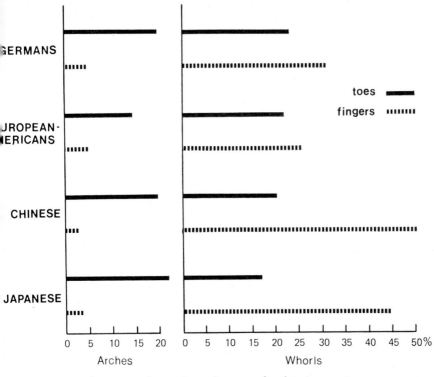

frequencies of arches and whorls on toes
compared with those on fingers

FIG. 5.   Frequencies of arches and whorls on toes in four racial samples, together with the corresponding frequencies on fingers. (Based on data from Steffens (1938), Newman (1936), Takeya (1933) and Hasebe (1918).)

Dankmeijer's arch/whorl index is, therefore, nearly always higher in females. This difference in pattern frequencies within a population makes it necessary for the sexes to be considered separately.

Inheritance of Dermal Ridge-Patterns

There is one great disadvantage in studies of pattern frequencies. We still do not know how the various pattern-types are inherited. From twin studies and other investigations there is little doubt that the arrangement of ridges into particular patterns has a hereditary basis. Much work on the subject, however, has led to inconclusive results, the general opinion being that the inheritance of pattern-types is very complex. This is true for finger-patterns (Elderton, 1920; Grüneberg, 1928; Böhmer and Harren, 1939; Essen-Möller, 1941), palmar patterns (Weninger, 1935, 1947; Weinand, 1937; Czik and Malan, 1938) and for sole patterns (Wichmann, 1956).

Although qualitative analyses of dermal ridge patterns have given inconclusive results, quantitative methods have proved more successful, at least in the case of finger patterns. Unfortunately, palmar patterns are more difficult to quantify and so quantitative methods have been used for features other than patterns on palms (Penrose, 1949b; Penrose, 1954; Pons, 1954). Quantitative genetical work has not yet been attempted on plantar patterns.

The Quantitative Approach—Finger-print Patterns

Certain difficulties complicate genetical work on finger patterns. Each individual has ten fingers and these often have different patterns. There is a marked tendency for some patterns to occur with greater frequency on some digits than others. Moreover, difficulties in classification are met in numerous cases, as a complete series of transitions exists between the three main pattern types, arches, loops and whorls. Classification is, therefore, to some extent subjective.

These difficulties can be overcome by using a quantitative measure of some feature of the pattern, instead of a qualitative classification. We owe such a method to Bonnevie (1924). She used the technique of ridge-counting for quantifying finger-print patterns. Ridge-counting was originally used by Galton (1895) as a means of sub-classifying loops for identification purposes. Bonnevie extended the method to all types of pattern.

The Method of Ridge-counting

The ridge-count, obtained from a rolled finger-print, is simply the number of ridges which cross or touch a straight line running from the triradius to the core or centre of the pattern (see Fig. 6). The triradius itself is not included in the count, nor is the final ridge, if it forms the centre of the pattern. In a simple arch, where there is no triradius, there is no count and the score is 0. In a loop there is one count, while in a typical whorl there are two counts, one for each triradius to the centre or centres of the pattern. In the latter case, only the higher count is used. The ridge-count gives an estimate of pattern-size. Although it is a somewhat arbitrary measurement, the ridge-count has various advantages. It is an objective feature of patterns and, in contrast with direct measurement of distance between triradius and core, it is independent of age. Moreover, it provides an excellent basis for quantitative analyses. Another advantage is that the ridge-counts on the ten fingers of an individual can be added to give a single value, the *total ridge-count*.

FIG. 6. The method of ridge-counting as applied to a loop. The white line drawn from the triradius to the core of the pattern shows the line of count. The ridge count is 13.

There is some agreement between ridge-count and pattern-type, but it is far from complete. Counts of 0 denote arches, either simple or tented, low counts usually indicate small loops, though occasionally small whorls occur, while high counts indicate either large loops or whorls. The relationship between ridge-count and pattern-type in 100 unrelated males (1000 fingers) and 100 unrelated females (1000 fingers) is shown in Fig. 7.

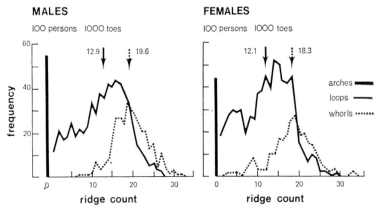

FIG. 7. The relation between pattern-size, as measured by ridge-count, and pattern-type. Data from 100 unrelated males and 100 unrelated females. The mean ridge-counts for loops and whorls in each sex are indicated by arrows.

Frequency Distributions of Total Ridge-count in the British Population

Before considering the genetics of pattern-size it was necessary to know something of the range and distribution of total ridge-count in the general (British) population. For this purpose the sexes were considered separately (Holt, 1949, 1955). The range was found to be from 0 to 285 ridges.

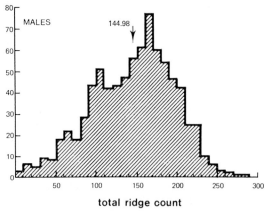

FIG. 8. Distribution of total ridge-count in a British population sample—825 males. The mean is indicated by an arrow. (Holt, 1955.)

Fig. 8 shows the frequency distribution for 825 males from a population sample. The distribution is non-Gaussian, being negatively skew and flattened. The value of g_1 (skewness), -0.290 ± 0.085, is highly significant, but the value of g_2 (kurtosis), -0.274 ± 0.171, only reaches 1.7 times its standard error. The presence of several peaks is a characteristic feature of the distribution. The mean ridge-count is 144.98, with a standard deviation of 51.08.

The frequency distribution of total ridge-count for 825 British females is given in Fig. 9. Again the distribution is negatively skew and flattened. The values of both g_1 (-0.230 ± 0.085) and g_2 (-0.451 ± 0.171) are highly significant. The mean is lower than in males, 127.27 ridges, and the standard deviation is 52.51.

The means for the two sexes are significantly different.

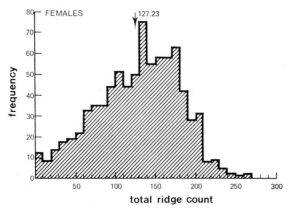

FIG. 9.   Distribution of total ridge-count in a British population sample—825 females. The mean is indicated by an arrow. (Holt,

The Inheritance of Total Finger Ridge-count

Statistical analyses of total ridge-count in family material have been made (Holt, 1952, 1956, 1957). In particular, familial correlations have been estimated so that the mathematical theory of Fisher (1918) and Penrose (1949a) on correlations between relatives could be used in determining the mode of inheritance. For this purpose data not only from parents and children, but from sibs and twins were necessary.

The estimated correlation coefficients for total ridge-count between the various types of relative are given in Table I. These results leave no doubt that finger-pattern size, measured by total ridge-count, is strongly determined by heredity. They also throw considerable light on the genetic process concerned.

In calculating all these correlations except the midparent-child, a sex correction was introduced, a value equal to the difference in mean between the sexes in the population sample being added to the total ridge-count of each female.

TABLE I
CORRELATIONS FOR TOTAL FINGER RIDGE-COUNT BETWEEN RELATIVES

	Correlation Coefficient	No. of pairs
Parent–Child	0·48	810 (200 families)
Midparent–Child	0·66 ± 0·03	405 ,,
Mother–Child	0·48 ± 0·04	405 ,,
Father–Child	0·49 ± 0·04	405 ,,
Parent–Parent	0·05 ± 0·07	200 ,,
Sib–Sib	0·50 ± 0·04	642 (290 sibships)
Monozygotic Twin–Twin	0·95 ± 0·01	80 ,,
Dizygotic Twin–Twin	0·49 ± 0·08	92 ,,

The parent-child correlation is very near a half. Note that the value of the interparental correlation (0·05 ±0·07) is less than its standard error and, therefore, not significant. It was necessary to know this value as positive correlation between parents increases all measurements of hereditary likeness.

The sib-sib correlation coefficient is also very nearly 0·5. Now according to Fisher (1918) the theoretical values for the correlations between parents and children and between sibs both have the value 0·5 when additive genes with independent effect but without dominance are present. Further, Penrose (1949a) has shown that the theoretical value of the midparent–child correlation coefficient is $1/\sqrt{2}$ or 0·71 under these conditions. The estimated midparent–child correlation is 0·66, a reasonably good approximation to the theoretical value.

Mother–child and father–child correlations are essentially the same. There is, therefore, no evidence here of a maternal (environmental) effect. Monozygotic twins are very highly correlated, the intraclass correlation based on data from 80 pairs is 0·95 ±0·01. This estimate is similar to that found by other workers (e.g. Lamy *et al.,* 1957). In contrast, the correlation between dizygotic twins, 0·49 ±0·08, is similar to that between ordinary sibs. The high monozygotic twin-twin correlation shows that the effect of environmental factors must be small, about 5 per cent of the total variability. The effect must be a maternal one, as only intra-uterine conditions can affect the formation and alignment of the ridges.

There is a further reason for supposing that the inheritance of total ridge-count is due to additive genes and that dominance is almost absent. The regression line test gives compatible results. With suitable data linear regression of offspring on the average value for the parents points to lack of dominance, and, therefore, to the presence of additive genes, while deviations from linearity are associated with dominance.

The regression of total ridge-count of child on midparent value is shown in Fig. 10. Data from 149 families, with 301 children were

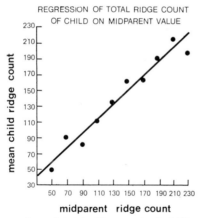

FIG. 10. Regression of total ridge-count of child on mid-parent value. The observed means and fitted straight regression line are shown.

used. The observed average child value for each of a series of midparent values is given, together with the fitted straight regression line. There is no significant deviation from linearity. The calculated regression coefficient is 0·92, agreeing with the theoretical value of one for a gene effect without dominance. (For this sample the midparent–child correlation coefficient was found to be 0·67 ±0·03 when no sex correction was used and 0·69 ±0·03 using a sex correction.)

From the frequency distributions for total ridge-count we can make certain assumptions concerning the number of genes having appreciable effect. The non-normality of the distributions in both sexes suggests that a comparatively small number of genes is involved. A large number of additive genes would give a nearly normal distribution (e.g. stature). Environmental effects and possibly the presence of minor modifying genes tend to blur the effects of the main genes. So far, it has not been possible to analyse the distributions into three components, corresponding with the three phenotypes produced by one pair of alleles, but this remains a tenable hypothesis.

Frequency Distributions of Total Ridge-Count in European Populations

As pattern-size measured by total ridge-count is an inherited metrical character, it should be of value in comparing different populations. So far we have little information concerning other races. There is, however, data comparable with the British from Portugal (da Cunha and Abreu, 1954), France (Lamy et al., 1956) and Sweden (Böök, 1957). The means and standard deviations for these four samples are given in Table II. All the distributions show the same essential features. While the French and Swedish distributions agree in being negatively skew and flattened, none of the values of g_1 and g_2 is significant. This is possibly due to the samples

TABLE II
MEANS AND STANDARD DEVIATIONS FOR TOTAL RIDGE-COUNT IN FOUR
EUROPEAN POPULATION SAMPLES

COUNTRY	MALES			FEMALES		
	No.	Mean	σ	No.	Mean	σ
Portugal (da Cunha and Abreu, 1954)	100	140·5	42·0	100	126·3	46·0
France (Lamy et al.,1956)	351	132·36	45·28	360	121·36	46·48
Sweden (Böök, 1957)	204	139·70	49·47	188	120·67	52·81
(Holt, 1955) Great Britain	825	144·98	51·08	825	127·23	52·51

being of smaller size than the British. No values of g_1 and g_2 are given for the Portuguese data.

Ridge-counts of Toe-print Patterns

Practically no work has been done on the ridge-counts of toe patterns. Steffens (1938) however, published data from the toes and fingers of 100 Germans, one member of each of 100 twin pairs. As was to be expected from the fact that arches are more frequent and whorls less frequent on toes than on fingers, the mean ridge-counts for all digits on feet were lower than the corresponding means on fingers.

Dermatoglyphic Patterns as Criteria for Racial Affinities

Cummins and Midlo (1943) have suggested that possibly "greater reliance in tracing racial affinities may be placed upon finger-prints" than on the dermatoglyphics of other regions. In the light of our present knowledge it seems that environmental factors play a lesser part in determining ridge arrangements on fingers than in determining those in other areas. Evidence for this conclusion is gained from qualitative comparisons of the patterns of monozygotic twin pairs and from quantitative genetical work on finger- and palm-prints.

It might be useful in summing up to see how dermatoglyphic patterns fulfil the requirements necessary for ascertaining racial relationships. Boyd (1940) listed five requirements for this purpose:

"1. The criteria must be objective.

2. They must not be subject to too much modification by the environment.

3. They must be determined by one, or a small number of genes.

4. They must be non-adaptive, that is, they must not have any great selective value in evolution.

5. They must not mutate at too high a rate."

With regard to the first criterion, qualitative classification of pattern-type is to some extent subjective. The ridge-count, however, is an objective feature of finger-print patterns—and so fulfils this requirement.

The second (that there must not be extensive environmental

modification) is completely satisfied by dermal ridge patterns. Only intra-uterine environment can affect the alignment of ridges and throughout post-natal life the patterns are environment-stable.

With respect to the third requirement (that the trait should have a simple, known mode of inheritance) dermatoglyphic patterns fall short of the ideal, in common with many other human traits. The genetics of pattern-type is complex and has not been elucidated. Now, however, we know something of the inheritance of total ridge-count; additive genes are involved and it seems likely that the number is small. We cannot yet pick out any of the genes concerned, so we know nothing about the mutation rates, criterion 5.

Concerning the fourth requirement (that the trait should not be affected by natural selection), dermal ridge patterns are presumably not affected by natural selection to any marked extent. In fact Galton (1892) speaking of finger patterns, stated: "Natural selection was shown to be inoperative in respect to individual varieties of pattern, and unable to exercise the slightest check upon their vagaries." In the case of total ridge-count the mathematical theory holds almost exactly because there is virtually no selection for this character.*

Dermatoglyphic patterns are, therefore, of considerable importance for anthropological investigations. They are among the few human traits unaffected by age. New approaches to the genetics of dermal ridges should prove fruitful in the future. In particular, the study of distortions in ridge arrangement produced by chromosome aberrations may well help to solve some of the problems of inheritance.

References

BÖHMER, K. and HARREN, F. (1939) Die Vererbung der Papillarlinien und ihre Bedeutung für den Nachweis der Vaterschaft. *Dtsch. Z. Ges. Gerichtl. Med.* **32**, 73.

BONNEVIE, K. (1924) Studies on papillary patterns of human fingers. *J. Genet.* **15**, 1.

BÖÖK, J. A. (1957) Frequency distributions of total finger ridge-counts in the Swedish population. *Hereditas* **43**, 381.

---

* There is no direct evidence of any action of natural selection, although it is suggested by the observed racial variation. It would, however, be possible to have selective forces which were too weak to be detected directly, but which were sufficient over a long period to account for the population differences in frequency of dermatoglyphic pattern-types, and also for the maintenance of the variability.

BOYD, W. C. (1940) Critique of methods of classifying mankind. *Amer. J. Phys. Anthrop.* **27**, 333.

CUMMINS, H. and MIDLO, C. (1943) *Finger Prints, Palms and Soles.* Philadelphia: Blakiston.

DA CUNHA, A. X. and ABREU, M. D. A. (1954) Impressoes digitais de Portuguese. *Contr. Antrop. Portug.* **5**, 315.

CZIK, J. and MALAN, M. (1938) Zur Erblichkeit der Hauptlinien und Muster der menschlichen Hand. *Z. KonstLehre* **21**, 186.

DANKMEIJER, J. (1934) De Beteekenis van Vingerafdrukken voor het anthropologisch Onderzoek, Dissertation, University of Utrecht. Utrecht, L. E. Bosch and Zoon. (Quoted by Cummins, H. and Midlo, C., 1943.)

DANKMEIJER, J. (1938) Some anthropological data on finger prints. *Amer. J. Phys. Anthrop.* **23**, 377.

ELDERTON, E. M. (1920) On the inheritance of the finger-print. *Biometrika* **13**, 57.

ESSEN-MÖLLER, E. (1941) Empirische Ähnlichkeitsdiagnose bei Zwillingen. *Hereditas* **27**, 1.

FISHER, R. A. (1918) The correlation between relatives on the supposition of Mendelian inheritance. *Trans. Roy. Soc. Edinb.* **52**, 399.

GALTON, F. (1892) *Finger Prints.* London: Macmillan.

GALTON, F. (1895) *Fingerprint Directories.* London: Macmillan.

GRÜNEBERG, H. (1928) Die Vererbung der menschlichen Tastfiguren. *Z. Indukt. Abstamm.-u. VererbLehre* **46**, 285.

HASEBE, K. (1918) Über das Hautleistensystem der Vola und Planta der Japaner und Aino. *Arb. Anat. Inst. Sendai* **I**, 13.

HOLT, S. B. (1949) A quantitative survey of the finger-prints of a small sample of the British population. *Ann. Eugen. (Lond.)* **14**, 329.

HOLT, S. B. (1952) Genetics of dermal ridges: inheritance of total finger ridge-count. *Ann. Eugen. (Lond.)* **17**, 140.

HOLT, S. B. (1955) Genetics of dermal ridges: frequency distributions of total finger ridge-count. *Ann. Hum. Genet (Lond.)* **20**, 159.

HOLT, S. B. (1956) Genetics of dermal ridges: parent-child correlations for total finger ridge-count. *Ann. Hum. Genet. (Lond.)* **20**, 270.

HOLT, S. B. (1957) Genetics of dermal ridges: sib-pair correlations for total finger ridge-count. *Ann. Hum. Genet. (Lond.)* **21**, 352.

LAMY, M., FRÉZAL, J., DE GROUCHY, J. and KELLY, J. (1957) Le nombre de dermatoglyphes dans un échantillon de jumeaux. *Ann. Hum. Genet. (Lond.)* **21**, 374.

NEWMAN, M. T. (1936) A comparative study of finger prints and toe prints. *Hum. Biol.* **8**, 531.

PENROSE, L. S. (1949a) *The Biology of Mental Defect.* Sidgwick and Jackson, London.

PENROSE, L. S. (1949b) Familial studies on palmar patterns in relation to mongolism. *Proc. 8th Int. Congr. Genet. (Hereditas,* supp. Vol.), 412.

PENROSE, L. S. (1954) The distal triradius *t* on the hands of parents and sibs of mongol imbeciles. *Ann. Hum. Genet. (Lond.)* **19**, 10.

PONS, J. (1954) Herencia de las líneas principales de la palma. Contribución a la Genética de los caracteres dermopapilares. *Trab. Inst. "Bernandos de Sahagun" de Antropología y Etnología del C.S.I.C.,* Barcelona **14**, 30.

SACHS, L. and BAT-MIRIAM, M. (1957) The genetics of Jewish populations. I. Finger print patterns in Jewish populations in Israel. *Amer. J. Hum. Genet.* **9**, 117.

STEFFENS, C. (1938) Über Zehenleisten bei Zwillingen. *Z. Morph. Anthr.* **37**, 218.

TAKEYA, S. (1933) Über die Hautleistenfigur der Zehen der Chinesen. *J. Orient. Med.* **19**, 36.

WEINAND, H. (1937) Familienuntersuchungen über den Hautleistenverlauf der Handfläche. *Z. Morph. Anthr.* **36,** 418.

WENINGER, M. (1935) Familienuntersuchungen über den Hautleistenverlauf am Thenar und am ersten Interdigitalballen der Palma. *Mitt Anthrop. Ges. Wien.* **55,** 182.

WENINGER, M. (1947) Zur Vererbung der Hautleistenmuster am Hypothenar der menschlichen Hand. *Mitt öst. Ges. Anthrop.* **73-77,** 55.

WICHMAN, D. (1956) Zur Genetik des Hautleistensystems der Fussohle. *Z. Morph. Anthr.* **47,** 331.

WILDER, H. H. (1904) Racial differences in palm and sole configuration. *Amer. Anthrop.* **6,** 244.

PIGMENTATION

Descriptive studies of normal population variation in pigmentation, quantified by the reflectance spectrophotometer, have continued, and there are now over 120 samples relating to indigenous populations in different parts of the world. Progress has been made with the elucidation of the inheritance of pigmentation, particularly in the Liverpool studies (Harrison and Owen 1964) and the several estimates of the numbers of genes responsible for the difference between African and European pigmentation are very similar (Stern 1970). Knowledge of the biochemistry of pigmentation variation has been further developed (Fitzpatrick *et al.*, 1967; Hempel 1968), but there has been little progress in understanding of the biological significance of variations.

ANTHROPOLOGISTS at one time paid considerable attention to human pigmentation but for some years its study, like that of many other quantitatively varying characters, has been neglected. There is evidence, however, of a return of interest in these characters and for this there are a number of reasons. In the first place anthropologists are no longer preoccupied with determining the evolutionary relations of populations but are concerned with every aspect of the nature of normal human differences. Secondly, both theoretical and practical advances have now made quantitative variation more amenable to genetic analysis; and thirdly, with the realization that most, if not all, population differences are adaptive, it is appreciated that much can be learned about characters whose mode of inheritance is not yet known because, after all, it is on the phenotype that selection directly acts.

By the standards of blood anthropology recent progress in the study of pigmentary differences has been very slow. But enough has been discovered to indicate the nature and biological significance of variation in human skin, hair and eye colour.

The Nature of Pigmentation

Edwards and Duntley (1939) first investigated systematically the nature of human skin colour by employing reflectance spectrophotometry. They found that although four types of pigment, melanin, melanoid, haemoglobin and carotenoid, and a scattering effect produced mainly by the stratum corneum were all involved in determining skin colour, it is variation in the melanin concentration that is principally responsible for interpopulation differences. In dark-skinned people the melanin is distributed throughout the epidermis and may also occur in considerable amounts in dermal cells, but in whites there is less and it is mainly confined to the stratum malpighi (Gates and Zimmermann, 1953). Melanin is formed solely by melanocytes which inject the particles of pigment into the basal epidermal cells (Billingham and Medawar, 1948). It is derived ultimately from tyrosine though, until recently, it was believed that dihydroxyphenylalanine was the precursor in skin. Fitzpatrick et al. (1950), however, have now demonstrated there is tyrosinase as well as dopa-oxidase activity in skin, at least following its ultra-violet irradiation. Striking evidence for an endocrine control has recently been discovered by Lerner and McGuire (1961). They found that the administration of melanocyte stimulating hormones, now extractable from mammalian pituitaries, profoundly darkened the skin colour of American Negroes. It seems that much of the pigmentary effect, formerly attributed to adrenocorticotrophin, is due to these melanocyte stimulating hormones.

The dependence of melanin formation on the amount of incident ultra-violet radiation is one of the most striking examples of human environmental lability. All people, apart from albinos, seem to possess the capacity to tan. Admittedly, Dr. J. J. T. Owen and I have found no difference in the skin colour of Negroes long resident in Liverpool and of those who had just arrived from Africa. This, however, was probably because we took our measurements of colour on the medial aspect of the upper arm. This site is not only often clothed and, anyway, protected by its position from exposure to sunlight, but also has a poor tanning capacity. It is therefore ideal for studying inherent differences, as it is so easily accessible.

The action spectrum for melanogenesis in tanning is the same as the erythemal spectrum. The maximum effect is produced by a wavelength of 295 mμ and the least effect, within the action spectrum, at 280 mμ. Wavelengths greater than 320 mμ do not induce "primary melanization" but a process of "pigment darkening" has been recognized which is brought about by light of longer wavelength, i.e. 420 mμ. This reaction which appears to be the conversion of a non-pigmented precursor of melanin to a pigmented state occurs within an hour of exposure to ultra-violet radiation, whereas 24 hr are required before primary melanization is evident (Blum, 1955a).

Except by artificial design, hair colour has little environmental lability, but melanin again is the principal pigment involved. There is, of course, no haemoglobin contribution—so important in determining skin colour—but the occurrence of specific red pigments introduces other complications. Little is as yet known about their chemistry; they may be related to melanin, but the work of Foster (1951) suggests that tryptophane might be a precursor. The observation of Sorby (1879) that red hair yields a pink pigment trichosiderin, when boiled with acid, has been followed up by Barnicot (1956a, 1956b) who has shown that a close correlation exists between the trichosiderin yield and the degree of redness. The melanin granules in hair have been investigated by Barnicot, Birbeck and Cuckow (1955) with the electron microscope which reveals the granules as being dense, elongated or rounded bodies between 0·5 μ and 1·0 μ in length and about 0·25 μ in breadth. It seems probable that the granules are larger in the darker hair shades. They appear to originate within the follicular melanocytes as small vacuoles in a distinctive area of peripheral cytoplasm. In albinos only small "ghosts" of granules occur, whilst in red hair melanocytes small granules containing discrete sub-particles are found (Birbeck, Mercer and Barnicot, 1956).

Little work has been done recently on the nature of eye colour, though it has long been known that variations in this character are due to the quantity and distribution of melanin. In the lighter eye shades the pigment is confined to the retinal layers of the iris, so that scattering effects are most marked.

Measuring Human Pigmentation

One of the reasons why the study of eye colour has lagged behind that of skin and hair is because there are still no objective methods of measuring the variation. Originally all these three characters were studied by methods involving visual matching with some artificial standard. Reflectance spectrophotometry has revolutionized the study of skin and hair colour, since it offers a means of objective measurement on a continuous scale, and the method has achieved wide applicability by the development of portable reflectometers which can be used for measuring at least skin colour in the field (Weiner, 1952). Reflectance spectrophotometry involves measuring the amount of light at different wavelengths reflected from a surface as compared with that reflected at the same wavelengths by a perfect white standard. It has been used by Sunderland (1956) for describing the geographical variation of hair colour in the United Kingdom, by Reed (1952) in his analysis of the genetics of red hair, by Barnicot (1958) and Lasker (1954) in field skin colour studies and by Harrison and Owen (1956) in our work on the hybrid populations in Liverpool. Many other comparable investigations are in progress.

The form of the skin reflectance curves of West Africans, Chinese, Europeans and some African-European hybrids are shown in Fig. 1. It is noteworthy that the lighter the skin colour the more marked is the characteristic haemoglobin absorption band at 545 mμ. Of importance in the genetic analysis of differences between Africans and Europeans is the disposition of the hybrid curves. To some extent these run diagonally across the graph being relatively nearer to the African values at the blue end of the spectrum than at the red.

Harrison and Owen (1956) have shown that information can be obtained from these reflectance values about the concentration of melanin. If a glass cell containing the pigment is placed on the arm of a blond European, and the concentration of melanin varied so as to give reflectance values similar to those obtained from Africans, Europeans and hybrids, the reciprocal of the reflectance values from the cell are linearly related to concentration. Actually, the relationship holds on various backgrounds so it is not likely to be affected by haemoglobin, as long as the contribution from this factor is constant. However, it is apparent from Fig. 2 that the linearity of the relationship extends over a wider range of melanin concentrations at the red end of the spectrum than in the blue and since, incidentally, variations in haemoglobin contribu-

filter

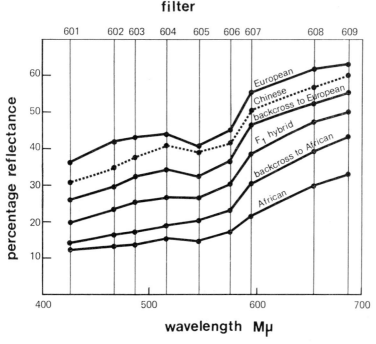

FIG. 1. Mean reflectance curves of Africans, Chinese, and Europeans and of some African-European hybrids [E.E.L. spectrophotometer].

tion have the smallest effect on reflectance of long wavelengths, there are good reasons for using the reflection of red light for comparative purposes. The complex optical properties of skin have not been taken into full account in these *in vitro* experiments but it seems unlikely that they will seriously affect the conclusions, since proportional linearity with concentration is still observed if a recently removed piece of epidermis is placed over the glass cell.

Skin reflectance studies have now been extended by the use of infra-red spectrophotometers and Jacquez *et al.* (1956) have investigated the effects of pigmentation on reflectance between 700 mμ and 2600 mμ. They found that white skin reflects most strongly at about 700 mμ and Negro skin at about 1000 mμ. Wavelengths longer than 1400 mμ are largely though not completely absorbed by both. In fact, between 1000 mμ and 2600 mμ it seems that Negroes absorb only about 2–3 per cent more radiation than Whites.

Variation in Heredity and Pigmentation

Quite apart from greying, there is a marked age effect on hair colour which particularly affects the lighter shades as late as up to thirty. On the other hand, though Garn, Selby and Crawford (1956a) found a slight lightening of the unexposed inner arm and

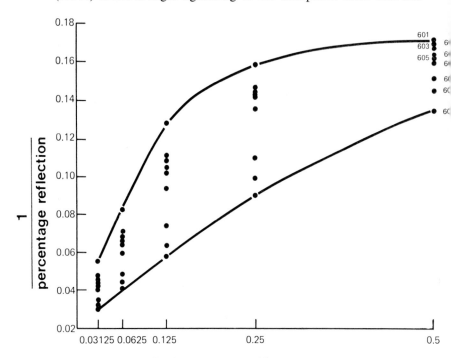

FIG. 2.   The relationship between melanin concentration and the reciprocal of the reflectance values at different wavelengths (the wavelengths of the different filters 601–609 are given in Fig. 1).

a darkening of the forehead with increasing age, and occasionally senescent changes are marked, skin as well as eye colour seem to alter little after the first few months of post-natal life. However, there is evidence for sex differences in both these characters (Boyd, 1950, and Barnicot, 1958). Owen and I have found that the skin of White female university students is lighter than that of male students, but this sex difference is not apparent among the poorer resident population of Liverpool who, incidentally, both adult and child, are slightly darker than the students. However, the possibility of differing cleanliness cannot be excluded as an explanation for this latter phenomenon.

The effects of pregnancy on areolar pigmentation have been quantified by Garn, Selby and Crawford (1956b).

Except for the occurrence of abnormal individuals such as albinos, hair and eye colour vary little except in peoples of European origin. Skin colour, however, shows a wide geographical variation in which a general pattern of decreasing pigmentation with increasing latitude is evident. Admittedly there are irregularities in this pattern. In Europe, for instance, the arctic zone peoples tend to be darker than the temperate zone ones. Further there are often quite marked local variations. In West Africa, for example, the Yorubas are lighter than their Ibo neighbours (Barnicot, 1958). Nevertheless, it is because a regular geographic pattern is discernible that skin colour was so frequently used for classifying races—a procedure not completely without justification, since many other anthropological characters display much the same general pattern of variation.

Compared with the differences between populations, the variation in skin colour within any one population is very small. In consequence, the little genetic research that has been done on this character has been concerned with the inheritance of inter-racial differences. Furthermore, most attention has been directed to the differences between Negroes and Whites, though Gates (1929) has obtained measurements on some other races. There is, of course, no reason to suppose that any of the genes involved in one racial difference are necessarily involved in another, even if the phenotypic differences are the same (Harrison, 1957). The first attempt to analyse the inheritance of Negro-White difference was made by Davenport and Danielson (1913) in Jamaica. They proposed a two allele system at two independent loci, with the genes responsible for increasing pigmentation having equal and additive effects. It has long been evident, however, that this was an inadequate system to account for the magnitude and nature of the observed variation. Gates (1949, 1953) has suggested that at least three loci are involved with genes of unequal effect, but his analysis suffers from the limitations of the colour chart method he has used for recording colour. More likely to be nearer the truth is Stern's conclusion (1953) that between four and six gene pairs are involved. Stern bases his conclusion on the comparison of the observed frequency distribution of skin colour in the American Negro, with model distributions of different numbers of gene pairs of equal and additive effect. Owen and I have investigated the problem with the use of a spectrophotometer in the hybrid families in Liverpool. This population is particularly suitable since the ancestry of many of the families is known back to the original miscegenation, but unfortunately backcross and F_2 generations are still far from common. We have, however, found that the genes, responsible for the colour differences, scale for additiveness very badly on the reflectance at 425 mμ but quite well on the reflectance at 545 mμ, and on the antilogarithms and reciprocals of the values at 685 mμ. It is not yet possible to decide which of these latter scales is best, but on each first-generation hybrid values deviate consistently from the mid-parent value in the direction of the White parent, indicating an overall potence of this genotype. Analyses of within family variation suggest that many loci are involved, but the data are as yet inadequate for firm conclusions.

The only recent contributions to the genetic study of hair colour are concerned with red hair. Neel (1943) found that the children of two red-haired parents are generally all red-haired and this has been confirmed by Reed (1952). But Reed has also shown that red hair cannot be distinguished as a discrete category in the population and that no clear cut segregation occurs in families. While the general evidence therefore points to recessiveness of the condition there are complications, one of which might well be occasional and variable heterozygous expression.

The Biological Significance of Skin Pigmentation

The very pattern of skin colour variation suggests the responsibility of some climatic factor as the selecting agent, and the individual's pigmentary lability to ultra-violet radiation strongly indicates that this particular factor is the one concerned. The amount of ultra-violet light an individual receives depends upon many atmospheric conditions as well as on latitude. Precise information on its local seasonal and diurnal variation is not available, but with one very notable exception it does seem that the distribution of pigmentary differences correlates well with intensity of ultra-violet radiation.

The Sudanese regions of Africa are exposed to the most intense solar radiation and here is found the heaviest pigmentation. Other hot desert and savanna peoples, such as the aboriginal Australian, who do not possess clothing or have acquired it only recently, tend also to be very dark skinned. Where less heavily pigmented peoples are living under comparable conditions, there is usually evidence that they are relatively recent immigrants—as, for example, in the hot deserts of the New World.

The clouded western seaboards of continents in temperate latitudes have the least sunlight. In the Arctic the open dust-free skies of summer and reflection from snow and ice expose the individual to strong ultra-violet radiation. As already mentioned, Arctic peoples tend to be darker than temperate ones.

It is well known that ultra-violet radiation with the same long wavelength limit as the erythemal and primary melanization action spectra, induces skin cancers in animals and there is considerable indirect evidence that basal-cell carcinomas in man may be caused by long exposure to strong sunlight. In White people, these skin cancers occur more frequently on the exposed parts of the body than in clothed regions, being in fact mainly confined to the face and the back of the hand. It also seems that they are more likely to develop in outdoor workers than in indoor ones. Furthermore, Dorn (1944) and Blum (1955) have shown there is a decrease in their incidence in American Whites with increasing latitude.

That melanin protects against the carcinogenic effects of ultra-violet light is suggested by the almost complete absence in Negroes of this type of carcinoma, for there is no evidence that the race is resistant to cancers in general. Blum (1955a), however, believes pigment to be unimportant in this role and that the thickness of the stratum corneum is the principal factor. He points out that increase in corneal thickness also follows exposure to ultra-violet radiation, that the erythemal threshold of skin incapable of synthesizing melanin is raised following such exposure and that the erythemal threshold of normal skin returns to its pre-irradiated level, while a tan is still evident. Nevertheless, Thomson (1955) has now shown that there is probably little or no difference in the thickness of the unexposed stratum corneum of Negroes and Whites, yet the opacity of the Negro corneum to ultra-violet radiation is much greater. This is not to deny that skin thickness performs an important protective function. The development of large horny pads on the back of the neck of African albinos suggest very strongly that it does, but at the same time this finding of Barnicot (1952) also indicates that melanin normally plays much more than a minor role.

It has been said that the selective significance of basal-cell carcinomas is very small, for they rarely cause severe debility and typically only develop late in life. This late development, however is no doubt partly due to the protection usually afforded by tanning and clothing and in any case can only be said to characterize the disease in temperate and sub-tropical zones. If naked Europeans were exposed to strong ultra-violet radiation, it seems likely that selection would strongly favour the darker individuals through their lower susceptibility to skin cancer. Nor must the immediate burning effect of solar radiation be neglected. Under natural condition severe sunburn will be extremely disabling.

The development of heavily pigmented skin in regions of strong solar radiation is not without its disadvantages. The difference between the incident radiation reflected by white and black skin obviously represents also a differential in that which is absorbed. Heer (1952) has calculated that Negro skin will absorb 34 per cent more energy than European skin when both are exposed to a blackbody radiating at 6000°K. Such a black body emits energy of similar spectral distribution to solar radiation at the surface of the earth's atmosphere.

The problem has been investigated experimentally by Baker (1958) who found that when nude Negro and White soldiers, matched in physique, exercised under desert conditions, there was a greater rise in rectal temperature and probably also in sweat production in the Negroes. This did not occur when the subjects were clothed, indicating that the differential solar heat load was responsible.

It seems, however, that this differential heat load is not as great as would be supposed solely from consideration of the comparative absorption and reflectance of black and white skin. While even the most penetrative infra-red radiation, i.e. at a wavelength of $1200m\mu$ is largely absorbed in the skin (Hardy and Muschenheim, 1936) it is to be expected that more is absorbed superficially in pigmented than in unpigmented epidermis. Certainly Wright (1958) found that the warmth sensation threshold can be lowered by painting the skin of a White individual with Indian ink. The surface temperature of Negro skin should, therefore, be greater than that of White when both are exposed to solar radiation and it is surprising that this has apparently never been verified experimentally. If such a difference does exist, it follows, under these circumstances, that dark-skinned individuals will gain less ambient heat than light-skinned ones when shade temperatures are above that of the body and lose more when they are lower.

While the development of heavy pigmentation in regions of strong insolation can be explained in terms of the protection this offers to the carcinogenic and burning effects of ultra-violet radiation, the unpigmented state that occurs particularly in peoples of European origin cannot be explained directly in the same terms. Unless this is an ancestral condition which now is without biological significance—an unlikely situation—there must be some positive advantage in being light skinned in regions of low ultra-violet incidence. The nature of this advantage is not certainly known but it is evident that ultra-violet radiation will have its greatest anti-rachitic activity when there is maximum penetration of the epidermis. Absence of melanin facilitates the penetration and in regions of low sunlight this factor could be critical (Coon, Garn, Birdsell, 1950).

Whether lightness of hair and eye colour in Europe is due directly to the depigmentation of skin is not clear. There is a loose association between the pigmentary state of all three structures, regions of very light skin being usually ones of blondism and blue eyes also. But it is evident that the condition of hair and eye colour is not merely a pleiotropic effect of skin colour genes.

It may be concluded that the pigmentation of peoples in most parts of the world can be explained as a compromise between the conflicting demands of protection from skin cancer and sunburn, thermoregulation and synthesis of Vitamin D. Such demands would tend to produce the observed phenoclines of skin colour, and in the New World where the first entry of man into the various climatic zones can be reasonably dated, it may well prove possible to evaluate their intensity.

There remains, however, the one very striking exception to the association between intensity of pigmentation and intensity of ultra-violet radiation. In equatorial and tropical forests according to data compiled by Richards (1952) little sunlight reaches ground level. In the Congo forests, for instance, the incident ultra-violet is as low as in Western Europe. This is not merely a function of the leaf canopy but also of the volcanic dust and water content of the atmosphere so that it tends to be true, at least now, of tree-cleared areas. Yet, excepting the savanna peoples the Congolese are as dark as any.

If the incident ultra-violet radiation is really as low as the data at present indicate and has long been so, there seem to be three possible explanations for the equatorial distribution of dark skin. First, it could be that heavy pigmentation is ancestral to the species, possibly at the time man's ancestors were haired, and no selection has subsequently operated to change it. Secondly, the inhabitants of the forests could be relatively recent immigrants from the neighbouring savannas. Or, thirdly, it is possible that some other quite different selective force is operating. Apart from being intellectually unsatisfying, the first two explanations do not seem to be adequate. The mere distribution of equatorial forest peoples suggests that some of them must have had a non-equatorial phase in their history. On the other hand, it seems very unlikely that recent immigration into the habitat accounts for the association, since at least four such immigrations would have to be hypothesized. Admittedly the Amazonian Indian has only relatively recently colonized his habitat but he is not as dark skinned as other forest peoples. The distribution of heavy pigmentation in S.E. Asia is the very opposite of that expected if it had recently been introduced. It looks much more as though an established dark-skinned group had been partly displaced by the ingression of lighter Mongoloid peoples from the north who themselves are now becoming darker. And there is much other evidence for this view.

Under these circumstances it seems reasonable to consider whether some other selective factor is operating on skin colour. One conceivable possibility which has been to some extent investigated is that a pigmented skin facilitates radiant heat loss. Radiation from a body is determined by the temperature and the emissivity coefficient of its surface. The effects of pigmentation on the skin temperature of an individual exposed to strong sunlight have already been considered; the question of importance here is whether or not the emissivity coefficient, or possibly the shaded skin temperature, is affected by the presence of epidermal melanin. Hardy and his colleagues (Hardy, 1934; Hardy and Muschenheim, 1934 and 1936; Hardy, Hammel and Murgatroyd, 1956) conclude from their extensive investigations that the human body, irrespective of its colour, acts as a near perfect black-body radiator.* Nevertheless, there are suggestions in Hardy and Muschenheim's 1934 data that small differences may exist in the emissivity coefficient of black and white skin. Although infra-red radiation longer than 1400 mμ is poorly reflected by human skin it is only completely absorbed if its wavelength is greater than about 8000 mμ. As already mentioned, spectroscopy has recently indicated that Negroes reflect some 2–3 per cent less than Europeans at near infra-red wavelengths longer than 1000 mμ (Jacquez et al., 1956). Differences of the same magnitude are reported by Hardy and Muschenheim but their data extend to the long wavelength limit of reflectance. The spectral range of heat emitted by the human body, having no overlap with the ground level spectrum of solar radiation, is from about 5000 mμ to about 20 000 mμ. Thus between 5000 and 8000 mμ where the body is not behaving as an absolutely perfect black body, one would expect a difference in emissivity of pigmented and unpigmented skin of about 2–3 per cent, since the emissive power of a body equals its absorbing power [Kirckhoff's Law]. That the differences may be yet greater is suggested by the work of Christiansen and Larsen (1935). They found that an arm congested with blood emitted more heat than an uncongested one with the same surface temperature. Criticisms have been levelled at the method of measuring the total radiation used by these workers and it is doubted whether the marked difference they found in the emissivity of skin and a perfect black

---

* Loss of radiant heat from such a body equals k ($T^4{}_S - T^4{}_E$) where k is the emissivity coefficient of a perfect black body and T_S and T_E are respectively the absolute temperature of the surface and the absolute temperature of the environment. (Stefan-Bolzmann Law.)

body is real, but there seems no reason to question the validity of their comparative findings on the arm.

There remains the possibility that the presence of melanin in skin facilitates the conductance of heat, thereby leading to a higher skin temperature for the same body temperature. As yet there appears to be no information for or against such a view.

By dealing at length with this possible relationship between skin colour and radiant heat loss I only wish to suggest that the problem is probably worthy of still further investigation. Certainly all the evidence indicates that any effect that may exist will be physically small, but this does not mean necessarily that the biological significance would also be small. In equatorial forests, where the atmosphere may be saturated for eighteen or more hours in the day, loss of heat by radiation must be extremely important in thermoregulation, and since the difference between environmental temperature and body temperature often does not exceed 10°F, quite small physical effects could have profound biological ones. It is becoming increasingly evident that when organisms are living at the limit of their heat tolerance virtually immeasurable environmental changes critically determine the chance of survival (Harrison, 1958). Nor is it relevant to argue that other factors such as linearity in physique or reduced metabolic rates are likely to have much more profound effects and by changing slightly could well negate colour differences. Selection operates on all the variation available until some equilibrium is reached and should pigmentation have any influence on heat loss, one would expect it to respond to environmental demands.

References

BAKER, P. T. (1958) Racial differences in heat tolerance. *Amer. J. Phys. Anthrop.* n.s. **16**, 287.

BARNICOT, N. A. (1952) Albinism in South-Western Nigeria. *Ann. Eugen. (Lond.)* **17**, 38.

BARNICOT, N. A. (1956a) The relation of the pigment trichosiderin to hair colour. *Ann. Hum. Genet.* **21**, 31.

BARNICOT, N. A. (1956b) The pigment, trichosiderin, from human red hair. *Nature (Lond.)* **177**, 528.

BARNICOT, N. A. (1958) Reflectometry of the skin in Southern Nigerians and in some Mulattoes. *Hum. Biol.* **30**, 150.

BARNICOT, N. A., BIRBECK, M. S. C. and CUCKOW, F. E. (1955) The electron microscopy of human hair pigments. *Ann. Hum. Genet.* **19**, 231.

BILLINGHAM, R. E. and MEDAWAR, P. B. (1948) Pigment spread and cell heredity in guinea-pigs' skin. *Heredity* **2**, 29.

BIRBECK, M. S. C., MERCER, E. H. and BARNICOT, N. A. (1956) The structure and formation of pigment granules in human hair. *Expt. Cell. Res.* **10**, 505.

BLUM, H. F. (1955a) *Sunburn Radiation Biology.* II, 487, Ed. Hollaender, A. McGraw-Hill, New York.

BLUM, H. F. (1955b) Ultra-violet radiation and cancer. *Radiation Biology.* II, 529. Ed. Hollaender, A. McGraw-Hill, New York

BOYD, W. C. (1950) *Genetics and the Races of Man.* Blackwell, Oxford.

CHRISTIANSEN, S. and LARSEN, T. (1935) On the heat radiating capacity of the human skin. *Skand. Arch. Physiol.* **72**, 11.

COON, C. S., GARN, S. M. and BIRDSELL, J. B. (1950) *Races.* Thomas, Illinois.

DAVENPORT, C. B. and DANIELSON, F. H. (1913) Skin color in Negro-White crosses. Publ. 188, Carnegie Institution of Washington.

DORN, H. F. (1944) Illness from cancer in the U.S.A. *U.S. Public Health Rept.* **59**.

EDWARDS, E. A. and DUNTLEY, S. Q. (1939) The pigments and colour of living human skin. *Amer. J. Anat.* **65**, 1.

FITZPATRICK, T. B., BECKER, S. W., LERNER, A. B. and MONTGOMERY, H. (1950) Tyrosinase in human skin: demonstration of its presence and its role in human melanin formation. *Science* **112**, 223.

FOSTER, M. (1951) Enzymatic studies on pigment-forming abilities in mouse skin. *J. Exp. Zool.* **117**, 211.

GARN, S. M. SELBY, S. and CRAWFORD, M. R. (1956a) Skin reflectance studies in children and adults. *Amer. J. Phys. Anthrop.*, n.s. **14**, 101.

GARN, S. M., SELBY, S. and CRAWFORD, M. R. (1956b) Skin reflectance during pregnancy. *Amer. J. Obstet. Gynec.* **72**, 974.

GATES, R. R. (1929) *Heredity in Man.* Constable, London.

GATES, R. R. (1949) *Pedigrees of Negro Families.* Blakiston, Philadelphia.

GATES, R. R. (1953) Studies of interracial crossing II. A new theory of skin colour inheritance. *Int. Anthrop. Ling. Rev.* **1**, 15.

GATES, R. R. and ZIMMERMANN, A. A. (1953) Comparison of skin color with melanin content. *J. Inv. Dermat.* **21**, 339.

HARRISON, G. A. (1957) The measurement and inheritance of skin colour in man. *Eug. Rev.* **49**, 2.

HARRISON, G. A. (1958) The adaptability of mice to high environmental temperatures. *J. Exp. Biol.* **35**, 892.

HARRISON, G. A. and OWEN, J. J. T. (1956) The application of spectrophotometry to the study of skin colour inheritance. *Acta Genet.* **6**, 481.

HARDY, J. D. (1934) The human skin as a black-body radiator. *J. Clin. Invest.* **13**, 615.

HARDY, J. D., HAMMEL, H. T. and MURGATROYD, D. (1956) Spectral transmittance and reflectance of excised human skin. *J. Appl. Physiol.* **9**, 257.

HARDY, J. D. and MUSCHENHEIM, C. (1934) The emission, reflection and transmission of infra-red radiation by human skin. *J. Clin. Invest.* **13**, 817.

HARDY, J. D. and MUSCHENHEIM, C. (1936) The transmission of infra-red radiation through skin. *J. Clin. Invest.* **15**, 1.

HEER, R. R. (1952) The absorption of human skin between 430 and 1010 mμ for black-body radiation at various colour temperatures. *Science* **115**, 15.

JACQUEZ, J. A., HUSS, J., MCKEEHAN, W., DIMITROFF, J. M. and KUPPENHEIM, H. F. (1956) Spectral reflectance of human skin in the region 0·7–2·6 μ. *J. Appl. Physiol.* **8**, 297.

LASKER, G. W. (1954) Photoelectric measurement of skin colour in a Mexican mestizo population. *Amer. J. Phys. Anthrop.*, n.s. **12**, 115.

LERNER, A. B. and MCGUIRE, J. S. (1961) Effect of alpha- and beta-melanocyte stimulating hormones on the skin colour of man. *Nature (Lond.)* **189**, 176.

NEEL, J. V. (1943) Concerning the inheritance of red hair. *J. Hered.* **34**, 93.

REED, T. E. (1952) Red hair colour as a genetical character. *Ann. Eugen. (Lond.)* **17**, 115.

RICHARDS, P. W. (1952) *The Tropical Rain Forest.* University Press, Cambridge.

SORBY, H. C. (1879) On the colouring matters found in human hair. *J. Roy. Anthrop. Inst.* **8**, 1.

STERN, C. (1953) Model estimates of the frequency of white and near-white segregants in the American Negro. *Acta Genet.* **4**, 281.

SUNDERLAND, E. (1956) Hair-colour variation in the United Kingdom. *Ann. Hum. Genet.* **20**, 312.

THOMSON, M. L. (1955) Relative efficiency of pigment and horny layer thickness in protecting skin of Europeans and Africans against solar ultra-radiation. *J. Physiol.* **127**, 236.

WEINER, J. S. (1952) A spectrophotometer for measurement of skin colour. *Man* **253**, 1.

WRIGHT, G. H. (1958) The effects of skin-blackening on warmth sensations and thresholds. *Clin. Sci.* **17**, 43.

References to Updating Paragraphs

ALSTROM, C. H. and LINDELIUS, R (1966) "A study of population movement in nine Swedish subpopulations in 1800–49". *Acta Genetica et Statistica Medica,* **16** (Supplement).

ALTER, M. (1967) "Dermatoglyphic analysis as a diagnostic tool". *Medicine,* **46,** 35–56.

ANTONINI, E. (1965) "Interrelationship between structure and function in hemoglobin and myoglobin". *Physiol. Rev.,* **45,** 123.

ARUNACHALAM, V. and OWEN, A. R. G. (1971) "Polymorphisms with linked loci". Chapman & Hall, London.

BIELICKI, T. and WELON, Z. (1964). "The operation of natural selection on human head form in an east European population". *Homo.,* **15,** 22–30.

BLACK, J. A. and DIXON, G. H. (1968). "Aminoacid sequence of the alpha chains of human haptoglobins" *Nature,* **218,** 736–41.

BOON, A. R. and ROBERTS, D. F. (1970) "The social impact of haemophilia". *J. Biosocial Science,* **2,** 237–64.

COHEN, B. H. (1970) "ABO and Rh incompatibility". *Amer. J. Hum. Gen.,* **22,** 412–440, 441–452.

DAVISON, B. C. C. (1965) "Epidermolysis bullosa". *J. Med. Gen.,* **2,** 233–242.

DAY, M. H. (1973). "Human Evolution". Taylor & Francis, London.

DEAN, G. (1971) "The porphyrias". Pitman, London.

DOBZHANSKY, T. (1970) "Genetics of the evolutionary process". Columbia Univ. Press, N.Y.

ELBUALY, M. S. and SCHINDELER, J. D. (1971) "Handbook of clinical dermatoglyphs". University of Miami Press, Florida.

FITZPATRICK, T. B., MIYAMOTO, M. and ISHIKAWA, K. (1967) "The evolution of concepts of melanin biology". *Arch. Dermatol.,* **96,** 305–323.

FRASER, G. R. (1972) "The short term reduction in birth incidence of recessive disease as a result of genetic counselling" *Human hered.,* **22,** 1–6.

FRIEDLANDER, J. G. (1971a) "Isolation by distance in Bougainville" *Proc. Nat. acad. Sci.,* **68,** 704–7.

FRIEDLANDER, J. G. (1971b) "The population structure of south-central Bougainville". *Am. J. Phys. Anth.,* **35,** 13–26.

FULLER, I. C. (1973) "Dermatoglyphics: a diagnostic aid?" *J. Med. Gen.,* **10,** 165–9.

GIBLETT, E. (1969) "Genetic markers in human blood". Blackwell, Oxford.

GLANVILLE, E. V. (1969) "Nasal shape, prognathism and adaptation in man". *Am. J. Phys. Anth.,* **30,** 29–38.

GOODMAN, M., BARNABAS, J., MATSUDA, G. and MOORE, G. W. (1971) "Molecular evolution in the descent of man". *Nature,* **233,** 604–613.

GRUBB, R. (1970). "The genetic markers of human immunoglobins". Chapman & Hall, London.

HARPENDING, H., and JENKINS, T. (1973) "Genetic distance among southern African populations" *in* CRAWFORD, M. H. and WORKMAN, P. L. (eds.) "Methods and theories of anthropological genetics" 177–199, Univ. New Mexico Press, Albuquerque.

HARRIS, H. (1970) "Principles of human biochemical genetics". North Holland, Amsterdam.

HARRISON, G. A. and OWEN, J. J. T. (1964) "Studies on the inheritance of human skin colour". *Annals Human Genetics,* **28,** 27–37.

HEMPEL, K. (1968) "Biosynthesis of melanin" *in* "The biologic effects of ultraviolet radiation" ed. F. URBACH, Pergamon, London. 305–314.

HOLLAND, J. H. (1973) "A brief discussion of the role of coadapted sets in the process of adaptation" *in* "Computer simulation in human population studies" ed. B. DYKE and J. W. MacCLUER. Academic Press, New York.

INGRAM, V. M. (1963). "The hemoglobins in genetics and evolution". Columbia, N.Y.

KIMURA, M. (1968) "Evolutionary rate at the molecular level". *Nature*, **217**, 624–6.

KING, J. L. and JUKES, T. H. (1969) "Non-Darwinian evolution" *Science*, **164**, 788–98.

LANGANEY, A., GOMILA, J. and BOULOUX, C. (1972). "Bedik: bioassay of kinship". *Hum. Biol.*, **44**, 475–8.

LANGLEY, C. H. and FITCH, W. M. (1973) "The constancy of evolution" *in* "Genetic structure of populations" ed. N. MORTON, 246–262. Hawaii Univ. Press.

LEHMANN, H. and CARRELL, R. W. (1969) "Variations in structure of human haemoglobin". *Brit. Med. Bull.*, **25**, 14–23.

LEHMANN, H. and HUNTSMAN, R. G. (1966) "Man's haemoglobins". North Holland, Amsterdam.

LIVINGSTONE, F. B. (1967). "Abnormal hemoglobins in human populations". Aldine, Chicago.

LIVINGSTONE, F. B. (1973) "Data on the abnormal hemoglobins and glucose–6–phosphate dehydrogenase deficiency in human populations". Tech. Report **3**, Museum of Anthopology. Univ. of Michigan.

MORAN, P. A. P. (1962) "The statistical processes of evolutionary theory". Clarendon Press, Oxford.

MORTON, N. E., YEE, S., HARRIS, D. E. and LEW, R. (1971). "Bioassay of kinship". *Theoretical population biology*, **2**, 507–24.

MOTULSKY, A. G. (1969) "Some evolutionary implications of biochemical variants in man". *Proc. 8th Internat. Cong. Anth. and Ethnol. Sci.*, **1**, 364–5.

MOTULSKY, A. G., FRASER, G. R. and FELSENSTEIN, J. (1971) "Public health and long term genetic implications of intrauterine diagnosis and selective abortion". *Birth Defects Orig. Art. Series*, **7**, 22–32.

NEEL, J. V. and SCHULL, W. J. (1968) "On some trends in understanding the genetics of man" *Perspectives Biol. Med.*, **11**, 565–602.

PEARSON, P. (1972) "The use of new staining techniques for human chromosome identification". *J. Med. Gen.*, **9**, 264–75.

PENROSE, L. S. (1967) "Fingerprint pattern and the sex chromosomes" *Lancet*, **1**, 198–200.

RACE, R. R. and SANGER, R. (1968) "Blood groups in man". Blackwell, Oxford.

REED, T. E. (1968) "Research on blood groups and selection from the Child Health and Development Studies, Oakland, California". *Amer. J. Hum. Gen.*, **20**, 129–141.

REED, T. E. (1974) "Selection and the blood group polymorphisms" (in press).

REED, T. E. and MILKOVICH, L. (1968) "The accuracy of blood grouping of cord blood specimens with special reference to the MN system". *Vox Sanguinis*, **14**, 9–17.

RIMOIN, D. L., MERIMEE, T. J., RABINOWITZ, D. McKUSIK, V. A. and CAVALLI-SFORZA, L. L. (1970) "Growth hormone in African Pygmies" *Lancet*, **2**, 523–6.

ROBERTS, D. F. (1973) "Climate and human variability". Addison Wesley, Reading, Mass.

SCHACHTER, H., MICHAELS, M. A., TILLEY, C. A. and CROOKSTON, M. C. (1973) "Qualitative differences in the N–acetyl–D–galactosaminyl transferases produced by human A_1 and A_2 genes". *Proc. Nat. acad. Sci.*, **70**, 220–4.

SHEPPARD, P. M. (1967) "Natural Selection and Heredity". Hutchinson, London.

SING, C. F., SHREFFLER, D. C., NEEL, J. V. and NAPIER, J. A. (1971) "Studies on genetic selection in a completely ascertained Caucasian population". *Am. J. Hum. Gen.,* **23**, 164–198.

SMITHIES, O., CONNELL, G. E. and DIXON, G. H. (1966) "Gene action in the human haptoglobins". *J. Mol. Biol.,* **21**, 213–224.

STEEGMAN, A. T. (1965) "Relationships between facial cold response and some variables of facial morphology". *Am. J. Phys. Anth.,* **23**, 355–62.

STERN, C. (1970) "Model estimates of the number of gene pairs in pigmentation variability of the Negro-Americans". *Hum. Hered.,* **20**, 165–168.

STEVENSON, A. C. and CHEESEMAN, E. A. (1956) "Hereditary deaf mutism, with particular reference to Northern Ireland". *Annals Human Genetics,* **20**, 117–231.

STEVENSON, A. C., JOHNSTON, H. A., STEWART, M. C. P., GOLDING, D. R. (1966) "Congenital malformations". *Bull. W.H.O.,* **34**, (Supplement).

SUTTON, H. E. (1970) "The haptoglobins" *in* "Progress in medical genetics" ed. A. G. STEINBERG and A. G. BEARN. **7**, 163–216.

TURNER, J. R. G. (1970) "Changes in mean fitness under natural selection" *in* "Mathematical topics in population genetics" ed K. KOJIMA, Springer-Verlag, Berlin.

VOGEL, F. (1970) "ABO blood groups and disease". *Am. J. Hum. Genetics,* **22**, 464–472.

WALKER, J. H., THOMAS, M. and RUSSELL, I. T. (1971) "Spina bifida and the parents". *Dev. Med. Child. Neurol.,* **13**, 462–476.

WEATHERALL, D. J. (1967) "The thalassemias" *in* "Progress in Medical Genetics" ed A. G. STEINBERG and A. G. BEARN, **5**, 8–57.

WELLS, R. S. and KERR, C. B. (1965) "Genetic classification of ichthyosis" *Arch. Dermatol.,* **92**, 1–6.

WILSON, A. C. and SARICH, V. M. (1969) "A molecular time scale for human evolution". *Proc. Nat. acad. Sci.,* **63**, 1088–1093.

AUTHOR INDEX

SUBJECT INDEX